Studies in Computational Intelligence

Volume 758

Series editor

Janusz Kacprzyk, Polish Academy of Sciences, Warsaw, Poland
e-mail: kacprzyk@ibspan.waw.pl

The series "Studies in Computational Intelligence" (SCI) publishes new developments and advances in the various areas of computational intelligence—quickly and with a high quality. The intent is to cover the theory, applications, and design methods of computational intelligence, as embedded in the fields of engineering, computer science, physics and life sciences, as well as the methodologies behind them. The series contains monographs, lecture notes and edited volumes in computational intelligence spanning the areas of neural networks, connectionist systems, genetic algorithms, evolutionary computation, artificial intelligence, cellular automata, self-organizing systems, soft computing, fuzzy systems, and hybrid intelligent systems. Of particular value to both the contributors and the readership are the short publication timeframe and the world-wide distribution, which enable both wide and rapid dissemination of research output.

More information about this series at http://www.springer.com/series/7092

László T. Kóczy · Jesús Medina
Editors

Interactions Between Computational Intelligence and Mathematics

 Springer

Editors
László T. Kóczy
Faculty of Engineering Sciences
Széchenyi István University
Gyõr
Hungary

and

Budapest University of Technology
  and Economics
Budapest
Hungary

Jesús Medina
Departamento de Matemáticas,
  Facultad de Ciencias
Universidad de Cádiz
Puerto Real, Cádiz
Spain

ISSN 1860-949X   ISSN 1860-9503   (electronic)
Studies in Computational Intelligence
ISBN 978-3-030-09054-8   ISBN 978-3-319-74681-4   (eBook)
https://doi.org/10.1007/978-3-319-74681-4

Printed on acid-free paper

This Springer imprint is published by Springer Nature
The registered company is Springer International Publishing AG
The registered company address is: Gewerbestrasse 11, 6330 Cham, Switzerland

Preface

Computational intelligence and traditional mathematics are two interwoven and extremely important scientific areas nowadays. For example, the treatment of big amount of data is an eminent goal in numerous topics which have been taken into consideration in important funding programs, such as the current European research and innovation program, Horizon 2020, which includes this problem in many of its subtopics. Hence, the interactions between computational intelligence and mathematics are very important in order to tackle a variety of important challenges. This volume is a new step in this direction including interesting contributions of engineers and mathematicians focused on the resolution of significant and actual computer science problems.

The first paper deals with a social network-related problem: the investigation of the two most up-to-date ranking algorithms (PageRank and Katz) used for identifying the most influential participants in the discussion on a particular topic. The research is based on a case study, namely, the "London Riots", using the Twitter database. In this study, most influential participating persons in the discussion are considered as trendsetters, i.e., having the ability to change the opinion of many others. The frequency of mentioning a participant's name and forwarding his/her tweets is used as the main measurable feature of a participant. After introducing the influence measures used by the two ranking algorithms mentioned above, the authors described their own experiments and results. They point out that for most users, the two methods agree but sometimes there are some surprising differences in the rank. While Katz uses simple frequency proportional mention weights, PageRank penalizes those mentions which come from less important participants. As the result of a deeper analysis, the authors conclude that PageRank has proved to be more robust a solution to identify influential users than the other top approaches.

The second paper deals with the problem of the Synthetic Aperture Radar (SAR) imagery filtering. The problem in SAR images is the presence of atmospheric conditions introducing noise and partial covering by clouds. In addition, a large amount of speckle, a multiplicative non-Gaussian noise degrades the quality of the SAR images. The study compares two classes of aggregation operators: the

Weighted Mean (WM) and the Weighted Ordered Average (OWA). A new family, merging the advantages of both previously mentioned classes, the Weighted OWA Operators (WOWA), offers a combination of advantages of the abovementioned two methods. The paper describes in detail the procedure of filtering the SAR images, with all three approaches. For learning the parameters, the authors use a Genetic Algorithm (GA), the classical evolutionary approach. As the GA is not one of the most efficient evolutionary methods, it is no surprise that the authors conclude that there is a strong dependence of the quality of the results on the number of generations. The final result points toward all proposed filters being useful, or even at the best—depending on the type of image and the circumstances of application. The directions of future research are indicated at the end.

The third paper proposes a novel combination of wavelet transform and fuzzy rule interpolation for evaluating very big data obtained from the telecommunication field. The engineering problem is the pre-evaluation of copper wire pairs from the point of view of transmission speed in SHDSL transmission networks. In the present best practice, the insertion loss must be measured at a large number of different frequencies for every wire pair. Based on several 100,000s of real measurement results, it is proposed that the combination of the Haar and Daubechies-4 wavelets combined with the Stabilized KH rule interpolation leads to a much better prediction of the transmission quality than any method known so far in the literature. The conclusion suggests that this new method might be useful in a much wider field of big data analysis, as well.

The fourth paper introduces a logic called the Molecular Interaction Logic, which semantically characterizes the Molecular Interaction Maps (MIM) and, moreover, makes it possible to apply deductive and abductive reasoning on MIMs in order to find inconsistencies, answer queries, and infer important properties about those networks. This logic can be applied to different metabolic networks, as the one given by the cancer, which can sometimes appear in a cell as a result of some pathology in a metabolic pathway.

The fifth paper is based on another logic. In this case in the general fuzzy logic programming called *multi-adjoint logic programming*. This framework was introduced by Medina et al. in 2001 as a general framework in which the minimal mathematical requirements are only considered in order to ensure the main properties given in the diverse usual logic programming frameworks, such as in possibilistic logic programming, monotonic and residuated logic programming, fuzzy logic programming, etc. Since its introduction, this framework has been widely developed from the theoretical and applied aspects by diverse authors. In this paper, the authors perform experiments which have shown the benefits of using a new *c*-unfolding transformation which reuses some variants of program transformation techniques based on unfolding, which have been largely exploited in the pure functional—not fuzzy—setting. Specifically, this new *c*-unfolding transformation has been applied to fuzzy connectives, and the authors have shown how to improve the efficiency of the proper unfolding process by reusing the very well-known concept of dependency graph. Furthermore, the paper includes a cost analysis and discussions on practical aspects.

The last two papers are in the area of the Formal Concept Analysis (FCA). This framework arose as a mathematical theory for qualitative data analysis and has become an interesting research topic both on its mathematical foundations and on its multiple applications. Specifically, the sixth paper provides an overview of different generalizations of formal concept analysis based on fuzzy sets. First of all, a common platform for early fuzzy approaches is included. Then, different recently fuzzy extensions have been recalled, such as the generalized extension given by Krajči, the multi-adjoint concept lattices introduced by Medina et al., heterogeneous extension studied by diverse research groups exemplified by those of Krajči, Medina et al., Pócs, Popescu, using heterogeneous one-sided extension given by Butka and Pócs, and the higher order extension presented by Krídlo et al. The paper also discusses connections between the related approaches.

The seventh paper is related to knowledge extraction from databases using rules, which is a compact tool in the representation of the knowledge. Based on concepts from lattice theory, this paper has introduced a new kind of attribute implications considering the fuzzy notions of support and confidence. The authors have also studied different properties of the particular case in which the set of attributes are intensions and, finally, they have also included an application to clustering for size reduction of concept lattices.

We would like to express our gratitude to all the authors for their interesting, novel, and inspiring contributions. Peer-reviewers also deserve our deep appreciation because their deep and valuable remarks and suggestions have considerably improved the contributions.

And last but not least, we wish to thank Dr. Tom Ditzinger, Dr. Leontina di Cecco, and Mr. Holger Schaepe for their dedication and help to implement and finish this large publication project on time maintaining the highest publication standards.

Győr, Budapest, Hungary   László T. Kóczy
Puerto Real, Cádiz, Spain   Jesús Medina

Contents

Page Rank Versus Katz: Is the Centrality Algorithm Choice Relevant to Measure User Influence in Twitter? 1
Hugo Rosa, Joao P. Carvalho, Ramon Astudillo and Fernando Batista

Weighted Means Based Filters for SAR Imagery 11
L. Torres, J. C. Becceneri, C. C. Freitas, S. J. S. Sant'Anna
and S. Sandri

On Combination of Wavelet Transformation and Stabilized KH Interpolation for Fuzzy Inferences Based on High Dimensional Sampled Functions 31
Ferenc Lilik, Szilvia Nagy and László T. Kóczy

Abductive Reasoning on Molecular Interaction Maps 43
Jean-Marc Alliot, Robert Demolombe, Luis Fariñas del Cerro,
Martín Diéguez and Naji Obeid

Efficient Unfolding of Fuzzy Connectives for Multi-adjoint Logic Programs 57
Pedro J. Morcillo and Ginés Moreno

On Fuzzy Generalizations of Concept Lattices 79
Lubomir Antoni, Stanislav Krajči and Ondrej Krídlo

Generating Fuzzy Attribute Rules Via Fuzzy Formal Concept Analysis 105
Valentín Liñeiro-Barea, Jesús Medina and Inmaculada Medina-Bulo

Page Rank Versus Katz: Is the Centrality Algorithm Choice Relevant to Measure User Influence in Twitter?

Hugo Rosa, Joao P. Carvalho, Ramon Astudillo and Fernando Batista

Abstract Microblogs, such as Twitter, have become an important socio-political analysis tool. One of the most important tasks in such analysis is the detection of relevant actors within a given topic through data mining, i.e., identifying who are the most influential participants discussing the topic. Even if there is no gold standard for such task, the adequacy of graph based centrality tools such as PageRank and Katz is well documented. In this paper, we present a case study based on a "London Riots" Twitter database, where we show that Katz is not as adequate for the task of important actors detection since it fails to detect what we refer to as "indirect gloating", the situation where an actor capitalizes on other actors referring to him.

Keywords Page Rank · Katz · User Influence · Twitter · Data Mining

1 Introduction

Nowadays, there are 288 million active users on Twitter and more than 500 million tweets are produced per day [17]. Through short messages, users can post about their

This work was supported by national funds through Fundação para a Ciência e a Tecnologia (FCT) under project PTDC/IVC-ESCT/4919/2012 and funds with reference UID/CEC/50021/2013.

H. Rosa (✉) · J. P. Carvalho · R. Astudillo · F. Batista
INESC-ID, Lisboa, Portugal
e-mail: hugo.rosa@insec-id.pt

R. Astudillo
e-mail: ramon.astudillo@insec-id.pt

J. P. Carvalho
Instituto Superior Técnico, Universidade de Lisboa, Lisboa, Portugal
e-mail: carvalho@insec-id.pt

F. Batista
ISCTE-IUL - Instituto Universitário de Lisboa, Lisbon, Portugal
e-mail: fmmb@insec-id.pt

feelings, important events and talk amongst each other. Twitter has become so much of a force to be reckoned with, that anybody from major brands and institutions, to celebrities and political figures use it to further assert their position and make their voice heard. The impact of Twitter on the Arab Spring [6] and how it beat the all news media to the announcement of Michael Jackson's death [15], are just a few examples of Twitter's role in society. When big events occur, it is common for users to post about it in such fashion, that it becomes a trending topic, all the while being unaware from where it stemmed or who made it relevant. The question we wish to answer is: "Which users were important in disseminating and discussing a given topic?"

Much like real life, some users carry more influence and authority than others. Determining user relevance is vital to help determine trend setters [16]. The user's relevance must take into account not only global metrics that include the user's level of activity within the social network, but also his impact in a given topic [18]. Empirically speaking, an influential person can be described as someone with the ability to change the opinion of many, in order to reflect his own. While [13] supports this statement, claiming that "a minority of users, called influentials, excel in persuading others", more modern approaches [4] seem to emphasize the importance of inter-personal relationships amongst ordinary users, reinforcing that people make choices based on the opinions of their peers.

In [2], three measures of influence were taken into account: "in-degree is the number of people who follow a user; re-tweets mean the number of times others forward a user's tweet; and mentions mean the number of times others mention a user's name". It concluded that while in-degree measure is useful to identify users who get a lot of attention, it "is not related to other important notions of influence such as engaging audience". Instead "it is more influential to have an active audience who re-tweets or mentions the user". In [8], the conclusion was made that within Twitter, "news outlets, regardless of follower count, influence large amounts of followers to republish their content to other users", while "celebrities with higher follower totals foster more conversation than provide retweetable content". The authors in [12] created a framework named "InfluenceTracker", that rates the impact of a Twitter account taking into consideration an Influence Metric, based on the ratio between the number of followers of a user and the users it follows, and the amount of recent activity of a given account. Much like [2], it also shows "that the number of followers a user has, is not sufficient to guarantee the maximum diffusion of information (…) because, these followers should not only be active Twitter users, but also have impact on the network".

In this paper, we analyze how two well known network analysis algorithms, PageRank and Katz, affect the computation of mention-based user influence in Twitter. Although these two methods have previously been compared [11] and found to have been equivalent, we show that the same conclusion does not apply in the context of social networks, and that PageRank is indeed more adequate. We base our conclusions on a real world case study of the 2011 London Riots, since it was an important social event where Twitter users were said to have played a role in its origin and dissemination.

2   User Influence Representation

We propose a graph representation of user's influence based on "mentions". Whenever a user is mentioned in a tweet's text, using the @*user* tag, a link is made from the creator of the tweet, to the mentioned user. For example, the tweet *"Do you think we can we get out of this financial crisis, @user B?"*, from @*user A*, creates the link: @*user A* ⟶ @*user B*. This is also true for re-tweets, e.g. the tweet *"RT @userC The crisis is everywhere!"* from @*user A*, creates the link: @*user A* ⟶ @*user C*.

This representation not only is an exact structural replica of the communication web between users, but it also provides dynamism to how influence can be given and taken across the graph.

In graph theory and network analysis, the concept of centrality refers to the identification of the most important vertices's within a graph, i.e., most important users. We therefore define a graph $G(V, E)$ where V is the set of users and E is the set of directed links between them.

3   Network Analysis Algorithms

The computation of user influence is done by applying a centrality based algorithm to the graph presented in Sect. 2. Here we present two of the most well-known and used centrality algorithms, Page Rank and Katz.

3.1   PageRank

Arguably the most well known centrality algorithm is PageRank [9]. It is one of Google's methods to its search engine and it was created as way for computing a ranking for every web page based on the graph of the web uses. In this algorithm, web pages are nodes, while back-links form the edges of the graph (Fig. 1). It is defined by Eq. 1 as $PR(v_i)$ of a page v_i.

$$PR_{vi} = \frac{1-d}{N} + d \sum_{v_j \in M(v_i)} \frac{PR(v_j)}{L(v_j)} \tag{1}$$

It can be intuitively said about Eq. 1, that a page has high rank if the sum of the ranks of its back-links is high. In it, v_j is the sum ranges over all pages that has a link to v_i, $L(v_j)$ is the number of outgoing links from v_j, N is the number of documents/nodes in the collection and d is the damping factor. The PageRank is considered to be a random walk model, because the weight of a page v_i is "the probability that a random walker (which continues to follow arbitrary links to move from page to page) will be at v_i at any given time. The damping factor corresponds to

Fig. 1 A and B are
back-links of C

the probability of the random walk to jump to an arbitrary page, rather than to follow
a link, on the Web. It is required to reduce the effects on the PageRank computation
of loops and dangling links in the Web" [11]. Dangling links are "simply links that
point to any page with no outgoing links (…) they affect the model because it is
not clear where their weight should be distributed" [9]. The true value that Google
uses for damping factor is unknown, but it has become common to use $d = 0.85$ in
the literature. A lower value of d implies that the graph's structure is less respected,
therefore making the "walker" more random and less strict.

3.2 Katz

Another well known method is the Katz algorithm [7]. It is a generalization of a
back-link counting method where the weight of each node is "determined by the
number of directed paths that ends in the page, where the influence of longer paths is
attenuated by a decay factor" and "the length of a path is defined to be the number of
edges it contains" [11]. It is defined by Eq. 2 "where $N(v_i, k)$ is the number of paths
of length k that starts at any page and ends at v_i and α is the decay factor. Solutions
for all the pages are guaranteed to exist as long as α is smaller than $\lambda > 1$, where
$1/\lambda$ is the maximum in-degree of any page" [11].

$$I_{vi} = \sum_{k=0}^{\infty} [\alpha^k N(v_i, k)] \qquad (2)$$

It was shown in [11] that "Katz status index may be considered a more general
form of PageRank because in can be modified, within a reasonable range, to be
equivalent to PageRank" and that under a "relaxed definition of equivalence (…)
PageRank and Katz status index is practically equivalent to each other" as long as
the number of outgoing links from any vertex is the same throughout the graph,
which is very unlikely for graph modeled from a social network. On the other hand,
"it is also possible to modify PageRank to become completely equivalent to Katz
status index", however, in that case, "the modified PageRank is no long a random
work model because it can no longer be modeled from a probabilistic standpoint"
[11].

4 Dataset

In order to test the network analysis methods presented above, a database from the London Riots in 2011 [3] was used. The London Riots of 2011 was an event that took place between the 6th and 11th August 2011, where thousands of people rioted in several boroughs of London with the resulting chaos generated looting, arson, and mass deployment of police. Although Twitter was said to be a communication tool for rioting groups to organize themselves, there is little evidence that it was used to promote illegal activities at the time, though it was useful for spreading word about subsequent events. According to [5], Twitter played a big role spreading the news about what was happening and "was a valuable tool for mobilizing support for the post-riot clean-up and for organizing specific clean-up activities". Therefore it constitutes a prime data sample to study how users exert influence in social networks, when confronted with such a high stakes event.

The Guardian Newspaper made public a list of tweets from 200 influential twitter users, which contains 17,795 riot related tweets and an overall dataset of 1,132,938 tweets. Using a Topic Detection algorithm [1], we obtained an additional 25,757 unhastagged tweets about the London Riots. It consists of a Twitter Topic Fuzzy Fingerprint algorithm [14] that provides a weighted rank of keywords for each topic in order to identify a smaller subset of tweets within scope. This method has proven to achieve better results than other well known classifiers in the context of detecting Topics within Twitter, while also being faster in execution. The sum of posting and mentioned users is 13,765 (vertices) and it has 19,993 different user mentions (edges), achieving a network connectivity ratio of $\frac{edges}{vertices} = 1.46$.

5 Experiments and Results

In this section, we compare the results of ranking the most influential users using Page Rank, Katz and a mentions based baseline. We proceed by performing an empirical analysis of the users in order to ascertain their degree of influence and their position in the ranks. The graphs and ranking were calculated using *Graph-Tool* [10].

Table 1 shows how both network analysis algorithms behave while highlighting the rank differences (shown by the arrows in the last column). A "Mentions rank" is used as a base line. Figure 2 provides a visual tool to the graph, as provided by PageRank.

There is an obvious relation between the number of mentions and the ranking provided by the application of both algorithms: the highest ranked users in either Katz and PageRank, are some of the most mentioned users in our dataset. In fact, the relation is more clear between Katz and the baseline Mentions based ranking: Table 1 shows that the rank in both approaches is always either identical (@guardian, @skynewsbreak, @gmpolice, etc...) or at most separated by two positions (@richardpbacon is ranked 27th based on mentions, and 29th based on Katz). In order to determine the relation

Table 1 London riots top 20 most influential users according to page rank, and comparison with Katz

User	Mentions		PageRank		Katz		
	#	Rank	Score	Rank	Score	Rank	
@guardian	160	2	0.0002854	1	0.022157	2	
@skynewsbreak	178	1	0.0002512	2	0.023479	1	
@gmpolice	122	4	0.0002128	3	0.019009	4	
@riotcleanup	107	6	0.0001767	4	0.017992	6	↗
@prodnose	67	14	0.0001761	5	0.014022	15	↗↗↗
@metpoliceuk	116	5	0.0001494	6	0.018709	5	
@marcreeves	69	11	0.0001476	7	0.014195	12	↗↗
@piersmorgan	78	8	0.0001465	8	0.014959	9	
@scdsoundsystem	69	12	0.0001442	9	0.014190	13	↗↗
@subedited	70	10	0.0001337	10	0.014278	11	
@youtube	48	20	0.0001257	11	0.012424	20	↗↗↗
@bbcnews	94	7	0.0001256	12	0.016426	8	↗↗
@mattkmoore	62	15	0.0001237	13	0.013614	16	↗
@richardpbacon	40	27	0.0001218	14	0.011771	29	↗↗↗
@lbc973	34	35	0.0001150	15	0.011432	34	↗↗↗↗
@skynews	74	9	0.0001113	16	0.014638	10	↗↗
@bengoldacre	61	17	0.0001055	17	0.013526	17	
@bbcnewsnight	68	13	0.0000988	18	0.014123	14	↗↗
@tom_watson	44	21	0.0000968	19	0.012107	22	↗
@paullewis	129	3	0.0000954	20	0.019602	3	↗↗↗
...							
@juliangbell	61	16	0.0000275	188	0.0166597	7	↗↗↗↗↗

The arrows indicate most relevant rank differences

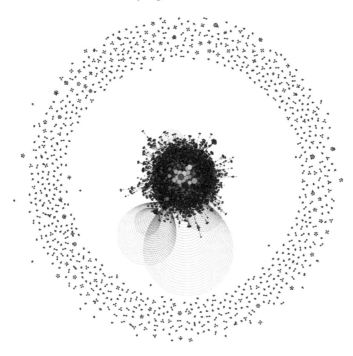

Fig. 2 User influence page rank graph - larger circles indicate larger user influence

between PageRank and "Mentions Rank", the Spearman correlation was calculated having achieved a value of $\rho = 0.9372$, which means they are heavily correlated. However, when limiting this calculation to the top 20, it changed to $\rho = 0.5535$, which implies that for the top users, just looking at the number of mentions, is not enough to determine influence.

An empirical analysis also shows that both Page Rank and Katz largely agree upon the ranking of most users, namely on the top two users: (i) @guardian, Twitter account of the world famous newspaper "The Guardian"; (ii) @skynewsbreak, Twitter account of the news team at Sky News TV channel. This outcome agrees with [8] previous statement, that, "news outlets, regardless of follower count, influence large amounts of followers to republish their content to other users" and can be justified by the incredibly high London Riots news coverage. Other users seem to fit the profile, namely @gmpoliceq, @bbcnews and @skynews. Most of the other users are either political figures, political commentators or jornalists (@marcreeves, @piersmorgan, and @mattkmoore).

However, when looking more closely at Page Rank versus Katz rankings, it is also possible to realize some notorious differences: Katz's third and seventh top ranked users are not in PageRank's top users. The reasons behind these differences in the ranking positions should be thoroughly analyzed since they could highlight the strengths and weaknesses of each algorithm in what concerns their capability to express user influence in social networks. The two cases end up being different and should be treated separately: (i) @paullewis, ranked 3rd by Katz shows up at 20th according to PageRank; (ii) @juliangbell, ranked 7th by Katz shows up at 188th according to PageRank.

The reason behind @paullewis high placement in the Katz rank is the number of mentions. As said previously, Katz is a generalization of a back-link counting method, which means the more back-links/mentions a user has, the higher it will be on the ranking. This user has 129 mentions, but PageRank penalizes it, because it is mentioned by least important users, which means a less sum weight is being transfered to it in the iterative process. This logic also applies to users @bbcnewsnight, @skynews and @bbcnews. Additionally, @paullewis is also an active mentioning user, having mentioned other users a total of 14 tweets, while @skynewsbreak and @guardian have mentioned none. As a consequence, Paul Lewis transfers its influence across the network while the others simply harvest it. There are several users that drop in ranking from PageRank to Katz for the very same reason. Users such as @prodnose, @marcreeves and @youtube do not have enough mentions for Katz to rank them higher.

User @juliangbell, despite mentioned often (61 times), is down on the PageRank because of indirect gloating, i.e., he retweets tweets that are mentioning himself: "*@LabourLocalGov #Ealing Riot Mtg: @juliangbell speech http://t.co/3BNW0q6*" was posted by @juliangbell himself. The user is posting somebody else's re-tweet of one of his tweets. As a consequence a link/edge was created from @juliangbell to @LabourLocalGov, but also from @juliangbell to himself, since his username is mentioned in his own tweet. Julian Bell is a political figure, making it acceptable that he would have a role in discussing the London Riots, but the self congratulatory behavior of re-tweeting other people's mentions of himself, is contradictory with the idea of disseminating the topic across the network. While Katz is not able to detect this effect, PageRank automatically corrects it, which is why, contrary to what is mentioned in previous works [11], it is our comprehension that Katz is not equivalent to PageRank in the task of detecting user relevance in social networks such as Twitter.

6 Conclusions

With this study, we have shown that in the context of user influence in Twitter, PageRank and Katz are not equal in performance, thus disproving previous claims. PageRank has proved a more robust solution to identify influential users in discussing and spreading a given relevant topic, specially when considering how it deals with indirect gloating, an item Katz fails to penalize.

References

1. Carvalho, J.P., Pedro, V., Batista, F.: Towards intelligent mining of public social networks' influence in society. In: IFSA World Congress and NAFIPS Annual Meeting (IFSA/NAFIPS), pp. 478 – 483. Edmonton, Canada (June 2013)
2. Cha, M., Haddadi, H., Benevenuto, F., Gummadi, K.P.: Measuring user influence in twitter: the million follower fallacy. In: In ICWSM '10: Proceedings of International AAAI Conference on Weblogs and Social (2010)
3. Crockett, K.S.R.: Twitter riot dataset (tw-short) (2011)
4. Domingos, P., Richardson, M.: Mining the network value of customers. In: Proceedings of the Seventh ACM SIGKDD International Conference on Knowledge Discovery and Data Mining, pp. 57–66. KDD '01, ACM, New York, USA (2001). http://doi.acm.org/10.1145/502512.502525
5. Guardian, T.: http://www.theguardian.com/uk/2011/dec/07/twitter-riots-how-news-spread
6. Huang, C.: Facebook and twitter key to arab spring uprisings: report. http://www.thenational.ae/news/uae-news/facebook-and-twitter-key-to-arab-spring-uprisings-report (June 2011). Accessed 02 May 2014
7. Katz, L.: A new status index derived from sociometric analysis. Psychometrika **18**(1), 39–43 (March 1953). http://ideas.repec.org/a/spr/psycho/v18y1953i1p39-43.html
8. Leavitt, A., Burchard, E., Fisher, D., Gilbert, S.: The Influentials: New Approaches for Analyzing Influence on Twitter (2009)
9. Page, L., Brin, S., Motwani, R., Winograd, T.: The Pagerank Citation Ranking: Bringing Order to the Web (1999)
10. Peixoto, T.: https://about.twitter.com/company
11. Phuoc, N.Q., Kim, S.R., Lee, H.K., Kim, H.: Pagerank vs. katz status index, a theoretical approach. In: Proceedings of the 2009 Fourth International Conference on Computer Sciences and Convergence Information Technology, pp. 1276–1279. ICCIT '09, IEEE Computer Society, Washington, DC, USA (2009). http://dx.doi.org/10.1109/ICCIT.2009.272
12. Razis, G., Anagnostopoulos, I.: Influencetracker: Rating the Impact of a Twitter Account. CoRR (2014). arXiv:1404.5239
13. Rogers, E.M.: Diffusion of Innovations (1962)
14. Rosa, H., Batista, F., Carvalho, J.P.: Twitter topic fuzzy fingerprints. In: WCCI2014, FUZZ-IEEE, 2014 IEEE World Congress on Computational Intelligence, International Conference on Fuzzy Systems, pp. 776–783. IEEE Xplorer, Beijing, China (July 2014)
15. Sankaranarayanan, J., Samet, H., Teitler, B.E., Lieberman, M.D., Sperling, J.: Twitterstand: News in tweets. In: Proceedings of the 17th ACM SIGSPATIAL International Conference on Advances in Geographic Information Systems, pp. 42–51. GIS '09, ACM, New York, NY, USA (2009). http://doi.acm.org/10.1145/1653771.1653781
16. Tinati, R., Carr, L., Hall, W., Bentwood, J.: Identifying communicator roles in twitter. In: Proceedings of the 21st International Conference Companion on World Wide Web, pp. 1161–1168. WWW '12 Companion, ACM, New York, NY, USA (2012). http://doi.acm.org/10.1145/2187980.2188256
17. Twitter: https://about.twitter.com/company
18. Weng, J., Lim, E.P., Jiang, J., He, Q.: Twitterrank: finding topic-sensitive influential twitterers. In: Proceedings of the Third ACM International Conference on Web Search and Data Mining, pp. 261–270. WSDM '10, ACM, New York, NY, USA (2010). http://doi.acm.org/10.1145/1718487.1718520

Weighted Means Based Filters for SAR Imagery

L. Torres, J. C. Becceneri, C. C. Freitas, S. J. S. Sant'Anna and S. Sandri

Abstract We address parameter learning for three families of filters for SAR imagery, based on WM, OWA and WOWA families of aggregation operators. The values in the weight vector associated to a WM filter correspond to the same positions in the input, whereas those in OWA filters consider the ordered positions of the input. WOWA operators make use of both vectors to weight an input data vector. Here we use Genetic Algorithms to learn the weight vectors for OWA, WM and WOWA filters and assess their use in reducing speckle in SAR imagery. We present an application using simulated images derived from a real-world scene and compare our results with those issued by a set of filters from the literature.

1 Introduction

Synthetic Aperture Radar (SAR) sensors are not as adversely affected by atmospheric conditions and the presence of clouds as optical sensors [13]. Moreover, they can be used at any time of day or night. The visual quality of SAR images is, however, degraded by sudden variations in image intensity with a salt and pepper pattern, due to the existence of a great amount of *speckle*, a multiplicative non-Gaussian noise, proportional to the intensity of the received signal [17]. Speckle in SAR image hampers the interpretation and analysis of this kind of image, as well as reduces the effectiveness of some image processing tasks such as segmentation and classification. For this reason, in order to process a SAR images, it is usually recommended to apply a filter before the segmentation or classification processes, even though the image quality with respect to some features may decrease, such as signal to noise ratio, spatial resolution, among others.

Filters for SAR imagery can be classified according to whether they take into account a model for speckle. A class of important representative of model-independent filters is the Ordered Statistical Filters (OSF) [2], based on order statistics [18].

L. Torres · J. C. Becceneri · C. C. Freitas · S. J. S. Sant'Anna · S. Sandri (✉)
Instituto Nacional de Pesquisas Espaciais (LAC/INPE), São José dos Campos, São
Paulo 12227-010, Brazil
e-mail: sandra.sandri@inpe.br

© Springer International Publishing AG 2018
L. T. Kóczy and J. Medina (eds.), *Interactions Between Computational Intelligence and Mathematics*, Studies in Computational Intelligence 758,
https://doi.org/10.1007/978-3-319-74681-4_2

Model-dependent filters are significantly more complex than the model-independent ones. Well-known representatives of this class of filters are the so-called Lee filter [10, 11] and its variations, such as the Refined Lee filter [12]. A more recent approach on model-dependent filters, called SDNLM (Stochastic Distances and Nonlocal Means) [27], is itself based on Buades et al.'s Nonlocal Means methodology [3].

Two important families of weighted aggregation operators are the Weighted Mean (WM) and the Weighted Ordered Average (OWA) [36]. In WM, each value in a weight vector **p** measures the importance of an information source with independence of the value that the source has captured, whereas in OWA, each value in a weight vector **w** measures the importance of a value (in relation to other values) with independence of the information source that has captured it [24].

OWA and WM filters are model-independent filters based on the use of OWA and WM operators, respectively. OWA filters are a particular representation of OSF, where, by definition, all coefficients are non-negative and sum up to 1. In [28, 29], we investigated the use of OWA filters to reduce speckle in SAR imagery and proposed strategies to learn vector **w** using Genetic Algorithms (GA) [6, 9]. In both works, we only dealt with intensity images; in [28] we addressed a single polarization (HH) and 3×3 windows, whereas in [29] we addressed filters for images in three polarizations (HH, HV and VV) using 5×5 windows.

The good results obtained by learning WM filters, that use vector **p**, and OWA filters, that use vector **w**, led us to investigate filters that would use both vectors **p** and **w**. Experiments that consisted in applying OWA and WM filters consecutively gave very poor results, so we turned our attention to families of operators that generalized both OWA and WM operators.

Weighted OWA operators (WOWA), proposed by Torra in 1997 [24], aims at taking advantage of both OWA and WM operators (see also [25, 26]). They have however an extra requirement, the choice of a particular type of function (called ϕ) to generate a vector weight ω, from **p** and **w**. We have first introduced the concept of WOWA filters in [30], and studied how the vectors **p** and **w** should be learned (sequentially or concomitantly), considering intensity images in a single polarization (HH). We also addressed learning WOWA filters for images obtained using the arithmetic mean of intensity images in polarizations HH, HV and VV [31].

In the present paper, we compare WM, OWA and WOWA filters, considering intensity images in three polarizations (HH, HV and VV), and study the influence of the number of generations in the GA on the performance of each filter. We report a series of experiments based on a fragment of a phantom described in [20]. To learn the WOWA weight vectors **p** and **w**, we use synthetic images in polarizations HH, HV and VV, that have been simulated using the parameters for Wishart distributions from a real-world scene, estimated in [23]. In order to assess the quality of the results produced by the GA, we use the Normalized Mean Square Error (NMSE) as fitness measure (see [1]). The results produced by these filters are compared to those issued by two model-dependent filters proposed in [12, 27].

This work is organized as follows. Section 2 discusses some basic aspects of SAR imagery. Section 3 describes WM, OWA and WOWA filters and Sect. 4 explains how

WOWA filters for SAR imagery are learned with the adopted GA. Section 5 presents an experiment in SAR imagery and Sect. 6 finally brings the conclusion.

2 Basic Concepts on SAR Imagery

SAR sensors operate in microwaves frequencies and generate images by measuring the energy reflected by targets on the earth's surface. SAR systems present side-looking geometry and are mounted on a platform which moves along a predefined trajectory, transmitting pulses at certain time intervals and receiving the backscattered signal. The backscattered signal is mainly affected by the target's electromagnetic characteristics and imaging geometry, as well as the particular features of the sensor system, such as frequency, polarization, incident angle, spatial resolution, etc. (see [19] for details on SAR systems).

Conventional imaging radars operate with a single, fixed-polarization antenna for both transmission and reception of radio frequencies signals [32]. In these systems, a single measure (scattering coefficient) is taken and is referred to as SAR data. Polarimetric radars usually operate employing horizontal (H) and vertical (V) linearly polarized signals for transmitting and/or receiving the energy.

The polarization combination for transmitted and received signals can lead to complex images, called polarimetric SAR data (PolSAR), which are also affected by speckle. These images are formed by a 2×2 scattering matrix, where the components are denoted by S_{HH} and S_{VV} (co-polarized data), and S_{HV} and S_{VH} (cross-polarized data). It is important to note that for reciprocal medium and monostatic systems, the components S_{HV} and S_{VH} are equal. Therefore, the four elements of the scattering matrix can be represented by a three-place vector as $[S_{HH} \quad S_{HV} \quad S_{VV}]$. Multiplying the vector $[S_{HH} \quad S_{HV} \quad S_{VV}]$ by its transposed conjugated vector $[S_{HH}^* \quad S_{HV}^* \quad S_{VV}^*]^t$, we obtain a 3×3 covariance matrix. The main diagonal elements, denoted by I_{HH}, I_{HV}, and I_{VV}, represent the intensity values of polarizations HH, HV and VV, respectively.

2.1 Filters for SAR Imagery

Given a window in an image, a filter simply substitutes the value of its central pixel by a function of the values of the pixels in the window. Two of the most basic filters use the arithmetic mean and the median as filtering function. In SAR imagery, the mean filter tends to reduce the speckle but it also tends to indiscriminately blur the image [14]. The median filter, on the other hand, reduces erratic variations by eliminating the lowest and highest pixel values [21].

Most filters employ the convolution operation. Given an image I, whose pixels take values in R, a $m \times m$ window around the central pixel (x, y) in I, and a matrix of coefficients $\gamma : \{-m, ..., 0, ..., m\}^2 \to R$, the result of convolution for (x, y) in the filtered image I_γ is calculated as

$$I_\gamma(x, y) = \sum_{i=-m,m} \sum_{j=-m,m} \gamma(i, j) \times I(x + i, y + j).$$

In Order Statistics Filters [2], the result of filtering for a given pixel is the linear combination of the ordered values of the pixels in the window around that pixel. They belong to the larger class of non-linear filters based on order statistics [18], being an application of L-estimators. An OSF is thus obtained when a convolution filter is applied on the ordered statistic of the pixel values in a window.

Some examples of model-independent filters are the directional means filters, in which only pixels in one of the twelve regions of the six orthogonal directions are considered (diagonals, rows and columns) [21] and the local region filters (see [22]), in which the window is divided in eight regions based on angular position, and the central pixel is replaced by the mean value of the subregion presenting the lowest variance.

The adoption of a model to the noise leads to more complex filters. One of such filters is the so-called Lee filter, in which speckle reduction is based on multiplicative noise model using the minimum mean-square error (MMSE) criterion [10, 11]. The Refined Lee filter [12], here called R-Lee filter, is an improved version of the Lee filter, and uses a methodology for selecting neighboring pixels with similar scattering characteristics.

Buades et al. [3] proposed a model-dependentt methodology that is well-suited for decreasing additive Gaussian noise, called Nonlocal Means (NL-means), which uses similarities between patches as the weights of a mean filter. The SDNLM (Stochastic Distances and Nonlocal Means) filter [27] is an adaptive nonlinear extension of the NL-means algorithm filter, in which overlapping samples are compared based on stochastic distances between distributions, and the p-values resulting from such comparisons are used to build the weights of an adaptive linear filter.

2.2 Image Quality Assessment for SAR Imagery

Assessing the performance of image filters is very hard [34]. Two important indices to measure the quality of filtered images are NMSE and SSIM, described in the following. Index NMSE (Normalized Mean Square Error) is a general purpose error measure, widely used in image processing (see [1]). Let r be the perfect information data and s an approximation of r; NMSE is calculated as:

$$\text{NMSE} = \frac{\sum_{j=1}^{n}(r_j - s_j)^2}{\sum_{j=1}^{n} r_j^2}, \tag{1}$$

where r_j and s_j refer to values in r and s at the same coordinates (the position of a given pixel in the case of images). NMSE always yield positive values, and the lower its value, the better is the approximation considered to be.

Index SSIM (Structural SIMilarity) measures the similarity between two scalar-valued images and can be viewed as a quality measure of one of them, when the other image is regarded as of perfect quality [35]. It is an improved version of the universal image quality index proposed proposed by [33]. This index takes into account three factors: (i) correlation between edges; (ii) brightness distortion; and (iii) distortion contrast. Let r and s be the perfect information and its approximation, respectively; SSIM is calculated as

$$\text{SSIM}(r, s) = \frac{\text{Cov}(r, s) + \alpha_1}{\widehat{\sigma}_r \widehat{\sigma}_s + \alpha_1} \times \frac{2\overline{r}\overline{s} + \alpha_2}{\overline{r}^2 + \overline{s}^2 + \alpha_2} \times \frac{2\widehat{\sigma}_r \widehat{\sigma}_s + \alpha_3}{\widehat{\sigma}_r^2 + \widehat{\sigma}_s^2 + \alpha_3}, \qquad (2)$$

where $\overline{r}$ and $\overline{s}$ are sample means, $\widehat{\sigma}_r^2$ and $\widehat{\sigma}_s^2$ are the sample variances, $\text{Cov}(r, s)$ is the sample covariance between r and s, and constants α_1, α_2 and α_3 are used the index stabilization. SSIM ranges in the $[-1, 1]$ interval, and the higher its value, the better is the approximation considered to be.

Some other measures that can be used to evaluate the quality of SAR imagery are, for instance, the equivalent number of looks (ENL), usually applied to intensity images in homogeneous areas, and index β_ρ, a correlation measure between the edges of two images (see [16]).

3 Filtering Based on WM, OWA and WOWA Operators

In the following, we first give the definitions of WM, OWA and WOWA families of operators. We then describe the filters obtained using these operators.

3.1 WM, OWA and WOWA Operators

Let $\mathbf{p}$ be a weighting vector of dimension n ($\mathbf{p} = [p_1 \; p_2 \; ... \; p_n]$), such that:

– (i) $p_i \in [0, 1]$;
– (ii) $\Sigma_i p_i = 1$.

A mapping $f_p^{wm} : R^n \to R$ is a Weighted Mean Operator (WM) of dimension n, associated to $\mathbf{p}$, if:

$$f_p^{wm}(a_1, ..., a_n) = \Sigma_i \; p_i \times a_i. \qquad (3)$$

Let $\mathbf{w}$ be a weighting vector of dimension n ($\mathbf{w} = [w_1 \; w_2 \; ... \; w_n]$), such that:

– (i) $w_i \in [0, 1]$;
– (ii) $\Sigma_i w_i = 1$.

A mapping $f_w^{owa} : R^n \to R$ is an Ordered Weighted Average Operator (OWA) of dimension n, associated to **w**, if [36]:

$$f_w^{owa}(a_1, ..., a_n) = \Sigma_i \, w_i \times a_{\sigma(i)}, \qquad (4)$$

where $\{\sigma(1), ..., \sigma(n)\}$ is a permutation of $\{1, ..., n\}$, such that $a_{\sigma(i-1)} \geq a_{\sigma(i)}$ for all $i = \{2,..., n\}$ (i.e., $a_{\sigma(i)}$ is the i-th largest element in $\{a_1, ..., a_n\}$).

Some well-known OWA operators are the mean, min, max and median, which are obtained with OWA vectors $w_{mean}, w_{min}, w_{max}$, and w_{med}, respectively. For instance, taking $n = 3$, we have: $w_{mean} = [1/3, 1/3, 1/3]$, $w_{min} = [0, 0, 1]$, $w_{max} = [1, 0, 0]$, and $w_{med} = [0, 1, 0]$.

The measures of *orness*, and *andness* [36], associated with a given vector **w** are defined as follows:

$$orness(w) = \frac{1}{n-1}\Sigma_i(n-i)w_i$$
$$andness(w) = 1 - orness(w)$$

Functions *orness* and *andness* describe an adjustment of the levels of "or" and "and", respectively, in the aggregation of a set of values.

Let **p** and **w** be weighting vectors as given above. A mapping $f_{w,p}^{wowa} : R^n \to R$ is a Weighted Ordered Weighted Average (WOWA) operator of dimension n, associated to **p** and **w**, if [24]:

$$f_{w,p,\phi}^{wowa}(a_1, ..., a_n) = \Sigma_i \, \omega_i \times a_{\sigma(i)}, \qquad (5)$$

where $\{\sigma(1), ..., \sigma(n)\}$ is a permutation of $\{1, ..., n\}$, for all $i = \{2,..., n\}$, such that $a_{\sigma(i-1)} \geq a_{\sigma(i)}$. Weight ω_i is defined as

$$\omega_i = \phi(P_\sigma(i)) - \phi(P_\sigma(i-1)), \qquad (6)$$
$$P_\sigma(i) = \Sigma_{j \leq i} \, p_{\sigma(j)}, \qquad (7)$$

and ϕ is a monotone increasing function that interpolates points $(0, 0)$ and $(i/n, \Sigma_{j \leq i} w_j)$, $i = 1, n$. Function ϕ is required to be a straight line when the points can be interpolated in a linear way. Torra [24] proves that the ω_i's compose a weighting vector of dimension n ($\omega = [\omega_1 ... \omega_n]$), such that:

– (i) $\omega_i \in [0, 1]$;
– (ii) $\Sigma_i \omega_i = 1$.

In [24], we can also find examples of non-linear functions to implement ϕ and the proof that OWA and WM are particular cases of WOWA operators.

WOWA operators, as well as OWA and WM operators, are particular cases of Choquet integrals [25, 26]. Even though more general, Choquet integrals have the inconvenient of requiring a large number of parameters, making them not practical for experiments involving learning. Indeed, if n is the number of data to be aggregated, OWA and the weighted mean require only n additional values and WOWA requires $2 \times n$, whereas Choquet integrals require 2^n values.

We now present a simple example to illustrate the use of OWA, WM and WOWA operators. Let $a_1 = 10$, $a_2 = 20$, and $a_3 = 0$. We thus have $\sigma(1) = 2, \sigma(2) = 1$, and $\sigma(3) = 3$. Therefore $a_{\sigma(1)} = 20$, $a_{\sigma(2)} = 10$, and $a_{\sigma(3)} = 0$.

Let $\mathbf{w} = [0.5\ 0.3\ 0.2]$ and $\mathbf{p} = [1/6\ 2/3\ 1/6]$. We thus obtain

- $f_w^{owa}(10, 20, 0) = 0.5 \times 20 + 0.3 \times 10 + 0.2 \times 0 = 13$,
- $f_p^{wm}(10, 20, 0) = 1/6 \times 10 + 4/6 \times 20 + 1/6 \times 0 = 15$.

Let ϕ_l be a linear by parts function, given by

$$\forall x \in (x_{k-1}, x_k], k \in \{1, n\}, \phi_l(x) = y_{k-1} + \frac{(x - x_{k-1})(y_k - y_{k-1})}{(x_k - x_{k-1})}, \quad (8)$$

where $\{(x_i, y_i)_{i=1,n}\}$ is a set of predetermined points. Using $\mathbf{w}$ to obtain the points $\{(x_i, y_i)_{i=1,n}\}$ as required in Eq. 6, and then applying them in Eq. 8, we obtain $\phi_l(0) = 0$, $\phi_l(1/3) = 0.5$, $\phi_l(2/3) = 0.8$ and $\phi_l(1) = 1$.

Taking σ and $\mathbf{p}$ above, we have $p_{\sigma(1)} = 2/3$, $p_{\sigma(2)} = 1/6$, and $p_{\sigma(3)} = 1/6$. We then obtain $P_{\sigma}(1) = 2/3$, $P_{\sigma}(2) = 5/6$, and $P_{\sigma}(3) = 1$, as well as $\phi_l(P_{\sigma}(1)) = 0.8$, $\phi_l(P_{\sigma}(2)) = 0.9$, and $\phi_l(P_{\sigma}(3)) = 1$. Therefore $\omega_1 = 0.8$, $\omega_2 = 0.1$, and $\omega_3 = 0.1$. Thus

- $f_{w,p,\phi}^{wowa}(10, 20, 0) = 0.8 \times 20 + 0.1 \times 10 + 0.1 \times 0 = 17$.

3.2 WM, OWA and WOWA Filters

OWA (respec. WM) filters (see [28, 29]) are obtained by applying OWA (respec. WM) weight vectors in the values inside a sliding window over a given image. Both WM and OWA filters are convolution filters, where coefficients γ are positive and add up to 1, with the coefficients in OWA being applied in the order statistic of the data, which makes them OSFs [7].

In the present paper, we introduce WOWA filters, based on a combination of WM and OWA filters. Given an image I, the application of a WOWA filter $F_{w,p,\phi}^{wowa}$ derives a filtered image $I_{w,p,\phi}^{wowa}$, as described below. OWA and WM filters F_w^{owa} and F_p^{wm} can be described similarly, but in a simplified manner.

Procedure $F_{w,p,\phi}^{wowa}(I)$

1. Transform a weight matrix $\mathbf{M}$ associated to a predefined neighbourhood, into a vector with n positions $\mathbf{p}$.
2. For each pixel in position (x, y) in the original image I, transform a window I' around (x, y), according to the predefined neighbourhood, into a vector of n positions $\mathbf{a}$.
3. Using $\mathbf{a}$, derive σ and $\mathbf{a}_\sigma$.
4. Using $\mathbf{w}$, $\mathbf{p}$ and ϕ, derive weight vector ω.
5. Calculate $f_{w,p,\phi}^{wowa}(a_1, ..., a_n)$.

6. Make the result become the value for position (x, y) in the filtered image:
   $I_{w,p,\phi}^{wowa}(x, y) = f_{w,p,\phi}^{wowa}(a_1, ..., a_n)$.

Given an image I and a position (x, y) in I, and considering a 3×3 window, in step 2 we derive vector $\mathbf{a}$, with 9 positions, as $(I(x - 1, y - 1), I(x - 1, y), I(x - 1, y + 1), I(x, y - 1), I(x, y), I(x, y + 1), I(x + 1, y - 1), I(x + 1, y), I(x + 1, y + 1))$. Vector $\mathbf{p}$ is derived in a similar way from matrix $\mathbf{M}$ in step 1.

4 Learning Weight-Based Filters for SAR Imagery

In the following, we describe our framework to learn the weights for OWA, WM and WOWA filters for SAR imagery. We first describe Genetic Algorithms, the optimization algorithm adopted in this work, and then how it is adapted to SAR imagery specificities.

4.1 Genetic Algorithms

Genetic Algorithms (GA), first proposed in [9] (see also [6]), combine Mendel's ideas about the codification of life in genes, with Darwin's ideas on the survival of the fittest (natural selection). They are search algorithms that evolve populations of candidate solutions, according to a fitness function that assesses the quality of these solutions to solve the problem at hand.

A candidate solution $c \in C = \{c_1, \ldots, c_k\}$ consists of a set of parameters to a function sol, that models the problem at hand. Each c can be thought of as a genotype (chromosome) and $sol(c)$ as its corresponding phenotype. A fitness function fit evaluates the candidate solutions; $fit(sol(c))$ should be proportional to the capacity of $c \in C$ in solving the problem at hand.

At each GA iteration, three processes (selection, crossover and mutation) take place, generating a new population C'. The selection process is such that the fittest candidates in C have a higher probability of being selected for reproduction. This process is usually performed by means of a roulette (the larger the fitness of an individual, the larger its share in the roulette wheel) or a set of tournaments (at each tournament, a set of individuals are chosen at random from the population and the winner is selected for reproduction). Different forms of elitism can also be used, by forcing the best candidates to be remain in the new population and/or to have a stronger influence on the creation of C'. The reproduction process, called crossover, creates two new candidate solutions by mixing the genotypes of two selected parent candidate solutions. In the mutation process, all new candidate solutions can suffer changes, according to a (usually small) probability, called the mutation rate, here denoted as ρ.

The first initial population is usually obtained at random, but in many applications, the use of a selected set of chromosomes may lead to better results. The stop criterion is usually a fixed number of iterations. The combination of selection, crossover and mutation provide GAs with a good equilibrium between exploration and exploitation of the search space.

In our work, each chromosome consists of weight vectors and each position in the chromosome contains a real number (a weight), that sum up to 1 altogether. The crossover operator adopted here consists in a linear combination of the parent chromosomes values. Given two parents c_1 and c_2, and random number $\alpha \in [0, 1]$, we generate two sons c_{12} and c_{21}, where $\forall i \in \{1, ..., n\}$, $c_{12}[i] = \alpha \times c_1[i] + (1 - \alpha) \times c_2[i]$ and $c_{21}[i] = (1 - \alpha) \times c_1[i] + \alpha \times c_2[i]$. The values in c_{12} and c_{21} are then normalized to guarantee that the weights sum up to 1.

In previous works, we tested a few strategies for mutation (see e.g. [28, 29]), and in the following we describe the ones that gave the best results, called A and B. In these strategies, the mutation rate is not applied on each position but on the chromosome as a whole. If a chromosome is selected for mutation, we randomly select the main position $1 \le q \le n$ in the chromosome to be changed, considering a uniform distribution. Then these mutation strategies differ as follows:

– A: The value in position q is multiplied by mutation rate ρ. The difference is divided by $n - 1$ and added to each of the remaining positions. Note that the larger is ρ, the larger is the change of the value in position q and the smaller is the change in the other positions in the vector.
– B: The value in q is increased with the value of its neighbour, considering the chromosome as a ring, and the neighbour receives value 0. When the neighbour to the right (respec. left) is considered, the strategy is named Br (respec. Bl).

In our previous works with OWA filters in SAR imagery, we verified that mutation operator A usually outperformed its B counterparts.

4.2   Learning SAR Filters with GAs

Given a SAR image on a given area, we need to have the means to assess the quality of filters applied on it. As done in previous works (see [28, 29]), here we adopt the following framework:

– samples from classes of interest are taken from a SAR image,
– for each class, the parameters of its associated distribution are estimated,
– using a phantom image, in which the regions are associated to the classes in the original image, a set of simulated images is created at random using the class distributions,
– the set of simulated images is partitioned in two sets, one for training and one for testing, and

– the best weight vectors found by the GA on the training set is used on the test set for evaluation.

Considering a $m \times m$ standard window, WM and OWA filters require chromosomes with $n = m^2$ positions to encode weight vectors $\mathbf{p}$ and $\mathbf{w}$, respectively. As for WOWA filters, when $\mathbf{p}$ and $\mathbf{w}$ are learned at the same time as here, each chromosome has $n = 2m^2$ positions.

In [29], we investigated the use of two strategies to learn OWA weight vectors for k polarizations, $k > 1$:

– S1: k GAs are run independently, one for each polarization, with weight vectors containing n positions.
– S2: a single GA is run, with the weight vector containing $k \times n$; when the weight vector is learned, it is split in k parts, resulting in a weight vector of n positions for each polarization.

The second strategy requires a more complex dealing with the GAs than the first one: even though selection, mutation and crossover remain basically the same, we have to ensure consistency, with each of the k parts of the chromosome adding up to 1. In our works, we verified that S2 produced better results for OWA filters in terms of quality of results and computational efficiency.

5 Experiments

We conducted a series of experiments based on images from an area in the Brazilian Amazon region, using L-band with wavelengths of [30 cm, 1 m] and frequencies of [1 MHz, 2 GHz] (see Fig. 1a). We took the parameters for Wishart distributions estimated in [23] for these images, with samples from 8 classes (see Fig. 1b), and used them on a fragment of a phantom described in [20] (see Fig. 1c), deriving 50 synthetic intensity images. Each simulated image has 240×240 pixels and was generated with 1-look.

Note that, when using a 3×3 window, weight vectors $\mathbf{p}$ and $\mathbf{w}$ have 9 positions each, and 25 positions with a 5×5 window. Here, we adopted strategy S2, i.e., weight vectors for the 3 polarizations are learned at the same time. Therefore, WM and OWA chromosomes have 27 and 75 positions for 3×3 and 5×5 windows, respectively. Here, vectors $\mathbf{p}$ and $\mathbf{w}$ are learned at the same time for WOWA filters (called strategy p+w in [30]). Therefore, the WOWA vectors have 54 and 150 positions for 3×3 and 5×5 windows, respectively.

We performed several experiments with different parametrizations. In particular, we verified that populations with 18, 36 and 72 elements generate quite similar results, and that the computational effort of using 72 elements is too large compared to the small gains it produces over its counterparts in some experiments. The use of different seeds also did not entail large differences in performance. However, we verified that a large number of generations does have an impact on performance for

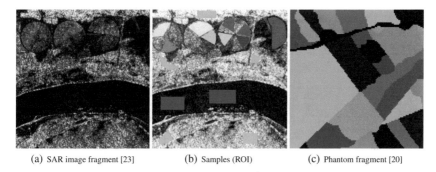

(a) SAR image fragment [23]   (b) Samples (ROI)   (c) Phantom fragment [20]

Fig. 1 SAR L-band false color composition, using HH (red), HV (green) and VV (blue) polarizations of the area of interest, ROIs and phantom used in the experiments

all filters. Here we report the use of OWA, WM and WOWA filters, considering the following parametrizations:

- selection type: roulette,
- number of generations: 30 and 120,
- population size: 36 elements,
- mutation rates: 0.2 and 0.8,
- seeds for random numbers: 2 and 271.

For each experiment, we performed a 5-fold cross-validation, using 40 images for training and 10 for testing in each fold. The elements in the initial population in each experiment were chosen at random. As fitness function for each fold in each parametrization, we took the means of the quality of the resulting filtered images, according to index NMSE (see Sect. 2.2).

Tables 1, 2 and 3 bring the results obtained for WM, OWA and WOWA filters, respectively, with parameters by GAs, according to NMSE, considering 5×5 windows. In the tables, we denoted the experiments as $\mathbf{E}/\xi/\psi$, where $\xi \in \{30, 120\}$ and $\psi \in \{0.2, 0.8\}$. Therefore, an experiment run with 30 generations and mutation rate 0.2 is denoted by $\mathbf{E}/\mathbf{30}/\mathbf{0.2}$. We mark in bold the best results obtained in each seed. The best aggregated results between the two seeds are marked with an asterisk ("*").

Table 1 NMSE mean (truncated) using WM on 50 images, with GA filters learned using 5 fold cross-validation, 5×5 windows, and mean NMSE of training images as fitness function

	Seed 2				Seed 271			
	HH	HV	VV	Agg	HH	HV	VV	Agg
E/30/.2	0.0547	0.0561	0.0570	0.0559	0.0540	0.0541	0.0555	0.0545
E/30/.8	0.0527	0.0535	0.0545	0.0536	0.0522	0.0532	0.0544	0.0533
E/120/.2	**0.0499**	**0.0508**	**0.0518**	**0.0508***	**0.0500**	**0.0508**	**0.0518**	**0.0509**
E/120/.8	0.0504	0.0511	0.0523	0.0513	0.0503	0.0512	0.0520	0.0512

Table 2 NMSE mean (truncated) using OWA on 50 images, with GA filters learned using 5 fold cross-validation, 5 × 5 windows, 36 elements in each population and mean NMSE of training images as fitness function

	Seed 2				Seed 271			
	HH	HV	VV	Agg	HH	HV	VV	Agg
E/30/.2	0.0515	0.0523	0.0536	0.0525	0.0518	0.0523	0.0537	0.0526
E/30/.8	0.0516	0.0523	0.0536	0.0525	0.0517	0.0524	0.0536	0.0526
E/120/.2	**0.0486**	**0.0494**	**0.0505**	**0.0495***	**0.0486**	**0.0494**	**0.0506**	0.0495*
E/120/.8	**0.0486**	0.0495	**0.0505**	**0.0495***	**0.0486**	**0.0494**	**0.0506**	**0.0495***

Table 3 NMSE mean (truncated) using WOWA on 50 images, with GA filters learned using 5 fold cross-validation, 5 × 5 windows, and mean NMSE of training images as fitness function

	Seed 2				Seed 271			
	HH	HV	VV	Agg	HH	HV	VV	Agg
E/30/.2	0.0515	0.0524	0.0538	0.0526	0.0515	0.0522	0.0536	0.0524
E/30/.8	0.0515	0.0523	0.0536	0.0525	0.0515	0.0523	0.0535	0.0524
E/120/.2	**0.0486**	0.0511	0.0512	0.0503	0.0493	0.0503	**0.0506**	**0.0501***
E/120/.8	0.0499	**0.0497**	**0.0511**	**0.0502**	0.0494	**0.0502**	0.0513	0.0503

In Table 1, we see that the mean NMSE of all results range between 0.0499 and 0.0570. The best aggregated result was obtained using 120 generations, seed 2 and mutation rate 0.2. We see in this table is that a large number of generations indeed increases the performance of WM filters.

In Table 2, we see that the mean NMSE of all results are more similar in the various parametrizations than in the case of WM, ranging between 0.0486 and 0.0537. The best aggregated result was obtained using 120 generations, seeds 2 and 271 and mutation rates 0.2 and 0.8. We see that also here, the most important factor on the results is the number of generations, with the seeds and mutation rates producing basically the same results, considering the individual polarizations as well as the aggregated results.

In Table 3, we see that the mean NMSE of all results are very close to those obtained by OWA filters, ranging between 0.0486 and 0.0538. The best aggregated result was obtained using 120 generations, seed 271 and mutation rate 0.2. Again here, the large number of generations produced the best results.

Comparing together the results of Tables 1, 2 and 3, we see that WM filters have been outperformed by both OWA and WOWA filters in all experiments. The results of OWA and WOWA are very similar; WOWA filters produced better results than OWA with a smaller number of generations, however, with a large number of generations, OWA produces the best overall results. In a nutshell, WOWA outperformed WM in our experiments but did not outperform OWA with our choice for function ϕ.

Table 4 brings the results obtained by the best aggregated GA-learned filters, considering the same type of window and the same number of folds. In Table 4, we also report the results for SDNLM and R-Lee filters, with the best parametrizations chosen after a few experiments, using 5×5 filtering window for both filters, with 3×3 patches, and significance level of 5% for for SDNLM, and with ENL = 1 for R-Lee. The best WM, OWA and WOWA filters for 5×5 windows are taken from Tables 1, 2 and 3, with ties solved considering non displayed decimal cases. For both 5×5 and 3×3 windows, the number of elements in the population is 36, and the best number of generations and best mutation rate are 120 and 0.2, respectively. Considering both types of windows, the seed value for OWA and WM is 2, whereas for WOWA the seed value is 271.

In Table 4, we see that the WOWA, OWA and WM filters, with weight vectors learned with a GA with 5×5 windows, outperformed all other filters considered here with respect to NMSE (used by the fitness function). In particular, OWA filters produced the best results, both in the individual polarizations as well as when they are aggregated. In what regards SSIM, the best performances were obtained with SDNLM and the mean filter. In the remaining of this section, we only discuss NMSE results.

Table 4 NMSE and SSIM mean (truncated) on 50 images, with GA filters learned using 5 fold cross-validation, 5×5 windows, and mean NMSE of training images as fitness function

	NMSE				SSIM			
	HH	HV	VV	Agg	HH	HV	VV	Agg
Unfiltered	0.997	0.074	1.025	1.008	0.068	0.084	0.053	0.068
SDNLM (5×5)	0.078	0.074	0.119	0.091	**0.166**	0.177	0.126	**0.157**
R-Lee (5×5)	0.079	0.075	0.125	0.093	0.155	0.167	0.124	0.149
OWA (5×5)	**0.048**	**0.049**	**0.050**	**0.049**	0.147	0.158	0.147	0.151
WOWA (5×5)	0.049	0.050	**0.050**	0.050	0.146	0.157	0.147	0.150
WM (5×5)	0.049	0.050	0.051	0.050	0.148	0.159	**0.148**	0.152
Mean (5×5)	0.064	0.063	0.124	0.084	**0.166**	**0.178**	0.124	0.156
Median (5×5)	0.153	0.153	0.203	0.170	0.123	0.131	0.099	0.118
Min (5×5)	0.927	0.928	0.934	0.930	0.001	0.001	0.000	0.000
Max (5×5)	9.754	9.868	9.590	9.737	0.024	0.026	0.019	0.023
OWA (3×3)	0.106	0.107	0.108	0.107	0.133	0.145	0.130	0.136
WOWA (3×3)	0.111	0.110	0.110	0.110	0.132	0.145	0.130	0.135
WM (3×3)	0.117	0.118	0.118	0.117	0.133	0.146	0.130	0.136
Mean (3×3)	0.115	0.116	0.116	0.116	0.137	0.150	0.133	0.140
Median (3×3)	0.189	0.190	0.190	0.190	0.107	0.116	0.104	0.109
Min (3×3)	0.807	0.807	0.808	0.807	0.007	0.007	0.007	0.007
Max (3×3)	4.969	5.025	4.971	4.988	0.039	0.046	0.037	0.040

In [28], we have verified that filters learned with 3×3 windows did not fare so well when compared with the model-based filters SDNLM and R-Lee, which use 5×5 windows, considering a GA configuration of 30 generations at most. That indicated that 5×5 windows are more appropriate 3×3 ones for our filters. Here, we see that with 120 generations, the model-based filters still outperform our filters with 3×3 windows but are outperformed when 5×5 windows are used. We see that even a very large increase in the number of generations in the GA is not capable of overcoming the negative impact that small windows cause on our filters.

Figure 2 brings an unfiltered synthetic image and the filtered images obtained from it using some of the methods considered here. We note that WM, OWA and WOWA

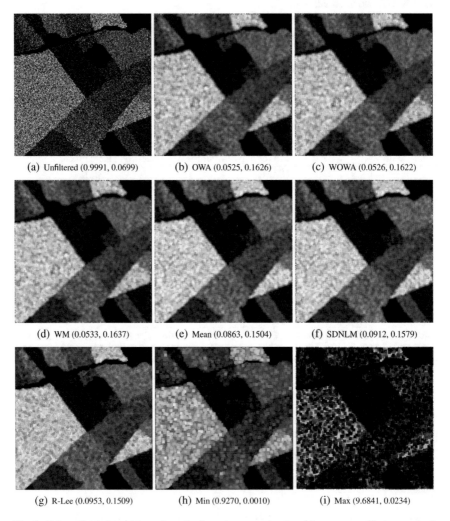

(a) Unfiltered (0.9991, 0.0699)   (b) OWA (0.0525, 0.1626)   (c) WOWA (0.0526, 0.1622)

(d) WM (0.0533, 0.1637)   (e) Mean (0.0863, 0.1504)   (f) SDNLM (0.0912, 0.1579)

(g) R-Lee (0.0953, 0.1509)   (h) Min (0.9270, 0.0010)   (i) Max (9.6841, 0.0234)

Fig. 2 False color composition of results from the same simulated images, considering methods using 5×5 windows, for HH (red), HV (green) and VV (blue) polarizations, with mean NMSE and SSIM from the polarizations inside parentheses

in Fig. 2 are visually superior to those obtained with both complex filters; SDNLM produces a blurred image and the Lee filter yields a pixelated image.

Figure 3 brings a square fragment of the false composition of the original image, with the ROIs from which the samples for each class were extracted (Fig. 3a),[1] the images obtained with OWA, WOWA and WM filters (Fig. 3b–d), all using as weights the best parametrization learned in fold 3, and the model-independent SDNLM and R-Lee filters (Fig. 3e, f, resp.).

Figure 3b, c, d show the effect of the OWA, WOWA and WM on windows of size 5×5 over the whole image. Albeit the noise reduction is evident, it is also clear that the blurring introduced eliminates useful information as, for instance, curvilinear details in the river area.

Figure 3f is the result of applying the R-Lee filter; although presenting a good performance, some details in the edges are eliminated, being worse than all the previous filters in this aspect. Figure 3e present the result of smoothing the original data set with the SDNLM filter at the level significance $\alpha = 10\%$. The noise effect is alleviated, e.g. the graininess is reduced specially in the planting and over the river areas, but fine details are more preserved than when the R-Lee filter is employed. Another important point is that land structures within the river are enhanced, and their appearance is maintained.

Figures 4 and 5 respectively illustrate the weight vectors **w** and **p** found in our best experiments. We see that the weights found are not distributed homogeneously (specially in what regards VV), but the the highest values for HH and HV are around the median. The orness found for vectors **w** for HH, HV and VV are respectively 0.494, 0.496 and 0.511, which is coherent with the visual inspection. The weights in vectors **p**, all polarizations considered, varied between 0.009 and 0.067. We see that HH and HV produce a more homogeneous distribution of weights than VV. Although there are large weight values in the borders, we see that the central pixel in all polarizations have a high weight value. We also conducted a set of experiments (not shown here) in which we learned a smaller number of weights that were repeated, according to the relative positions to the central pixel for **p**, and to the median for **w**, but the results were inferior to those obtained here.

The GA was run on a machine with the following specifications: Intel i7, CPU 2.60 GHz, RAM with 16 GB, Windows 10, Fortran with Force 2.0 compiler. Considering 5 folds, with 10 images in each fold, 36 elements in the population, and 30 generations, the GA processing for all our filters (WM, OWA and WOWA) took approximately 30 min for each polarization for 3×3 windows and approximately 1 h for 5×5 windows. When the number of generations is multiplied by 4 (120), the computation times also increase 4 times, in all cases.

---

[1] The available image with ROIs use the Pauli composition, that takes the complex part of the images into account. For this reason, the colours are different from the filtered images, that use the same composition used in Fig. 1a, which only takes the intensity images into account.

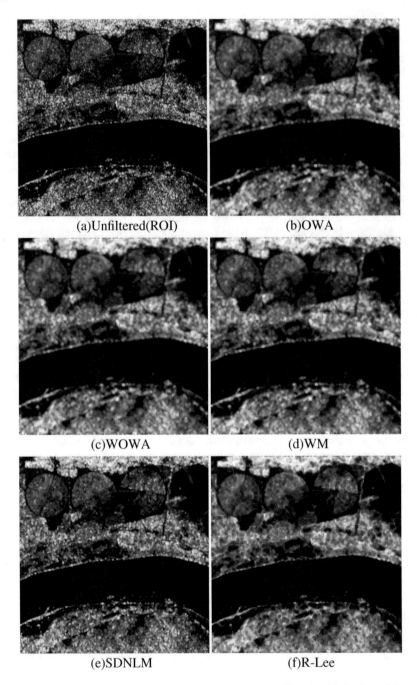

Fig. 3 False color composition of the original SAR image unfiltered, with ROIs, and images obtained with a set of filters

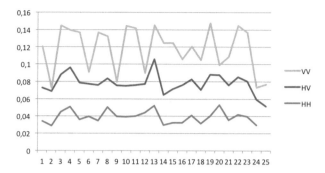

Fig. 4 Weight vectors **w** found in the best experiment

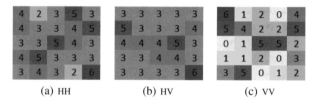

(a) HH (b) HV (c) VV

Fig. 5 Weight vectors **p** found in the best experiment (values multiplied by 10^2 with a single digit displayed)

6 Conclusions and Future Work

In this paper, we addressed OWA, WM and WOWA filters, based on OWA, WM and WOWA operators (see [25, 36]), respectively. We compared their ability to reduce speckle in SAR imagery, considering simulated images from three polarizations (HH, HV and VV) and two types of windows (3×3 and 5×5). The weight vectors of the three filters were learned with Genetic Algorithms, for both a small and a large number of generations.

Our experiments have shown that the number of generations used in the G.A. has a strong impact on the quality of the results. They have also shown that the size of the windows is however the most crucial factor for obtaining good results.

We compared the results against a set of filters from the literature, including model-independent filters WM and OWA (which also had their parameters learned) and two model-dependent ones, SDNLM [27], a recent parametrized family of filters (here with, parameters chosen after trial-and-error), and the well-known Refined Lee filter [12]. OWA filters, followed by WOWA and WM filters, with parameters learned using 5×5 windows outperformed all other filters, according to indice NMSE, used in the learning process.

The quality of filters depend to what one wants to extract from the filtered images. Visual inspection indicates that the sole use of one index to assess the quality of filters may not be enough for some applications. We are currently investigating the use multi-objective optimization to produce filters that take into account not only the error, measured by NMSE, but the preservation of the structures in the image, using a measure based on edge-detection.

This work and previous ones show that automatic learning of parameters can be crucial for parametrized families of filters. For instance, the filters obtained using the min, max, median and arithmetic mean operators are all particular cases of the OWA and WOWA operators and fared significantly worse than their learned counterparts. To be fair, the results obtained by OWA and WOWA filters should also be compared to model-dependent filters whose parametrizations are also learned, but this is out of the scope of this work.

As future research, we intend to use other alternatives to model function ϕ, used by the WOWA operators, instead of the linear by parts function adopted here. We also intend to study the influence of initial populations in the GA, and the use of other types of windows, such as the ones based on non-standard neighbourhoods (see, e.g. [5]). Last but not least, we intend to use the whole information contained in a POLSAR image, taking into account also the complex components, and not only the real components, as we have done so far.

Acknowledgements The authors are indebted to Vicenç Torra and to Lluis Godo for discussions on WOWA operators, and to Benicio Carvalho for providing computational resources.

References

1. Baxter, R., Seibert, M.: Synthetic aperture radar image coding. MIT Lincoln Lab. J. **11**(2), 121–158 (1998)
2. Bovik, A.C., Huang, T.S., Munson, D.C.: A generalization of median filtering using linear combinations of order statistics. IEEE Trans. ASSP **31**(6), 1342–1349 (2005)
3. Buades, A., Coll, B., Morel, J.M.: A review of image denoising algorithms, with a new one. Multiscale Model. Simul. **4**(2), 490–530 (2005)
4. Carlsson, C., Fullér, R.: Fuzzy reasoning in decision making and optimization. In: Studies in Fuzziness and Soft Computing Series. Springer (2002)
5. Fu, X., You, H., Fu, K.: A statistical approach to detect edges in SAR images based on square successive difference of averages. IEEE GRSL **9**(6), 1094–1098 (2012)
6. Goldberg, D.E.: Genetic Algorithms in Search, Optimization, and Machine Learning. Addison-Wesley (1989)
7. Grabisch, M., Schmitt, M.: Mathematical morphology, order filters and fuzzy logic. In: Proceedings of FuzzIEEE 1995, Yokoham, pp. 2103–2108 vol. 4 (1995)
8. Herrera, F.: Genetic fuzzy systems: status, critical considerations and future directions. Int. J. Comput. Intell. Res. **1**(1), 59–67 (2005)
9. Holland, J.H.: Adaptation in Natural and Artificial Systems. University of Michigan Press, USA (1975)
10. Lee, J.-S., Grunes, M.R., de Grandi, G.: Polarimetric SAR speckle filtering and its implication for classification. IEEE Trans. GRS **37**(5), 2363–2373 (1999)
11. Lee, J.-S., Grunes, M.R., Mango, S.A.: Speckle reduction in multipolarization, multifrequency SAR imagery. IEEE Trans. GRS **29**(4), 535–544 (1991)
12. Lee, J.-S., Grunes, M.R., Schuler, D.L., Pottier, E., Ferro-Famil, L.: Scattering-model-based speckle filtering of polarimetric SAR data. IEEE Trans. GRS **44**(1), 176–187 (2006)
13. Lee, J.-S., Pottier, E.: Polarimetric radar imaging: from basics to applications. In: Optical Science and Engineering. Taylor and Francis, UK (2009)

14. Mascarenhas, N.: An overview of speckle noise filtering in sar images. In: 1st Latin-American Seminar on Radar Remote Sensing—Image Processing Techniques, pp. 71–79 (1997)
15. Mittal, A., Moorthy, A.K., Bovik, A.C.: No-reference image quality assessment in the spatial domain. IEEE Trans. Image Process. **21**(12), 4695–4708 (2012)
16. Moschetti, E., Palacio, M.G., Picco, M., Bustos, O.H., Frery, A.C.: On the use of Lee's protocol for speckle-reducing techniques. Lat. Am. Appl. Res. **36**(2), 115–121 (2006)
17. Mott, H.: Remote Sensing with Polarimetric Radar. Wiley, USA (2006)
18. Pitas, I., Venetsanopoulos, A.N.: Nonlinear Digital Filters: Principles and Applications. Springer, Berlin (2013)
19. Richards, J.A.: Remote Sens. Imaging Radar. Springer, Signals and Communication Technology Series (2009)
20. Saldanha, M.F.S.: Um segmentador multinível para imagens SAR polarimétricas baseado na distribuição Wishart. Ph.D. Thesis, INPE, Brazil (2013)
21. SantAnna, S.J.S., Mascarenhas, N.: Comparação do desempenho de filtros redutores de "speckle" . In: VIII SBSR, pp. 871–877 (1996)
22. Sheng, Y., Xia, Z.G.: A comprehensive evaluation of filters for radar speckle suppression. In: IGARSS '96, vol. 3, pp. 1559 – 1561 (1996)
23. Silva, W.B., Freitas, C.C., Sant'Anna, S.J.S., Frery, A.C.: Classification of segments in PolSAR imagery by minimum stochastic distances between Wishart distributions. IEEE J-STARS **6**(3), 1263–1273 (2013)
24. Torra, V.: The weighted OWA operator. Int. J. Intell. Syst. **12**(2), 153–166 (1997)
25. Torra, V.: On some relationships between the WOWA operator and the Choquet integral. In: Proceedings of IPMU'98, Paris, France (1998)
26. Torra, V., Narukawa, Y.: Modeling Decisions: Information Fusion and Aggregation Operators. Springer, Berlin (2007)
27. Torres, L., Sant'Anna, S.J.S., Freitas, C.C., Frery, A.C.: Speckle reduction in polarimetric SAR imagery with stochastic distances and nonlocal means. Pattern Recogn. **47**(1), 141–157 (2014)
28. Torres, L., Becceneri, J.C., Freitas, C.C., Sant'Anna, S.J.S., Sandri, S.: OWA filters for SAR imagery. In: Proceedings of LA-CCI'15, Curitiba, Brazil, pp. 1–6 (2015)
29. Torres, L., Becceneri, J.C., Freitas, C.C., Sant'Anna, S.J.S., Sandri, S.: Learning OWA filters parameters for SAR imagery with multiple polarizations. In: Yang, X.-S., Papa, J.P. (eds.) Bio-Inspired Computation and Applications in Image Processing, pp. 269–284. Elsevier, Netherlands (2016)
30. Torres, L., Becceneri, J.C., Freitas, C.C., Sant'Anna, S.J.S., Sandri, S.: WOWA image filters. In: Proceedings of CBSF'16, Campinas, Brazil (2016)
31. Torres, L., Becceneri, J.C., Freitas, C.C., Sant'Anna, S.J.S., Sandri, S.: Comparing OWA and WOWA filters in mean SAR images. In: Proceedings of SBSR'17, S.J.Campos, Brazil (2017)
32. Ulaby, F.T., Elachi, C.: Radar Polarimetry for Geoscience Applications. Artech House, USA (1990)
33. Wang, Z., Bovik, A.C.: A universal image quality index. IEEE SPL **9**(3), 81–84 (2002)
34. Wang, Z., Bovik, A.C., Lu, L.: Why is image quality assessment so difficult? In: IEEE ICASSP, vol. 4, pp. 3313–3316, Orlando (2002)
35. Wang, Z., Bovik, A.C., Sheikh, H.R., Simoncelli, E.P.: Image quality assessment: from error visibility to structural similarity. IEEE Trans. Image Process. **13**(4), 600–612 (2004)
36. Yager, R.R.: On ordered weighted averaging aggregation operators in multi-criteria decision making. IEEE Trans. Syst. Man Cybern. **18**, 183–190 (1988)

On Combination of Wavelet Transformation and Stabilized KH Interpolation for Fuzzy Inferences Based on High Dimensional Sampled Functions

Ferenc Lilik, Szilvia Nagy and László T. Kóczy

Abstract A new approach for inference based on treating sampled functions is presented. Sampled functions can be transformed into only a few points by wavelet analysis, thus the complete function is represented by these several discrete points. The finiteness of the teaching samples and the resulting sparse rule bases can be handled by fuzzy rule interpolation methods, like KH interpolation. Using SHDSL transmission performance prediction as an example, the simplification of inference problems based on large, sampled vectors by wavelet transformation and fuzzy rule interpolation applied on these vectors are introduced in this paper.

Keywords Fuzzy inference · Performance prediction · Fuzzy rule interpolation
Wavelet analysis

1 Introduction

Due to the great number of input values, making inference on phenomena which can be described by large-sized vectors is difficult and expensive. In order to construct efficient inference systems, simplification of the input space is needed. This simplification makes the process of the inference easier, however, it unavoidably rises the system's level of uncertainty and inaccuracy. During our previous research on performance prediction of physical links of telecommunications access networks, we had to encounter such problems in two ways.

Due to the limited calculation capacity not all the measured values can be used as the bases of the inference. Drastically lowering the number of the measured frequency dependent input values caused an inaccuracy in the final results. Later, this type of sparseness will be referred to as vertical sparseness.

F. Lilik (✉) · S. Nagy · L. T. Kóczy
Széchenyi István University, Győr 9026, Hungary
e-mail: lilikf@sze.hu

L. T. Kóczy
Budapest University of Technology and Economics, Budapest 1117, Hungary

L. T. Kóczy and J. Medina (eds.), *Interactions Between Computational Intelligence and Mathematics*, Studies in Computational Intelligence 758,
https://doi.org/10.1007/978-3-319-74681-4_3

As it is not possible to measure all possible data, the finite teaching sample set will naturally result in sparse rule bases: there will be points, which will be outside of all the supports of the antecedent sets of the corresponding dimensions. Later, this behaviour will be referred to as horizontal sparseness.

In Sect. 2 the primary technical problem underlying the research on performance prediction is briefly reviewed, the first version of the inference method, its test results and the horizontal and vertical sparseness derived from the simplification of the input space are also shown. In Sect. 3 wavelet transformation and fuzzy rule interpolation as the algorithmic techniques applied in a combined way for handling the problems of simplification are described, and in Sect. 4 we present the test results of the new approach based on these techniques.

2 Fuzzy Performance Prediction of Physical Links in Telecommunications Access Networks of SHDSL Transmission

In certain cases, the technological demands of modern telecommunications services can be hardly, or can not be fulfilled by the existing telecommunication networks. This is especially true for mobile or copper wire pairs based access networks. Moreover, the performance possibilities of the individual links of these networks can be totally different from one another. Therefore, lately, telecommunications service providers apply various, usually rather resource demanding methods for predicting the performance of the individual physical links. The aim of our research was constructing a simple, but acceptably precise performance predicting method.

In this work, as an example, we have used symmetrical copper wire based access networks with SHDSL connections. SHDSL [1] is a symmetrical digital telecommunications transmission method, which establishes high speed data transmission over symmetrical wire pairs. Although the newest members of the SHDSL systems can reach higher data transmission speed (bit rate) than the one studied in the present contribution, our approach can be successfully applied also for those systems.

In this context, performance is a property of the physical link, therefore it has to be predicted from the measurable physical parameters of the link. Based on the related ITU-T recommendations [1, 2] and on our own measurements and investigations, SHDSL performance can be predicted from the frequency dependent insertion loss of the studied line [3]. Insertion loss is theoretically a continuous function in the lowest 1.5 MHz wide frequency domain, where the SHDSL transmission is realized. For technical reasons, only discrete values were measured from 10 kHz in 10 kHz wide steps, thus resulting in insertion loss values of the studied wire pairs at 150 discrete frequency points. The available maximal data transfer rate of the lines were also measured by installing SHDSL equipments to the bare lines. The measured insertion loss series were clustered into 5 groups according to the measured bit rates. This grouping is similar to the practice of telecommunication service providers when offering packages of DSL services.

Based on insertion loss values at 6 well-selected characteristic frequencies, fuzzy rule bases were created to predict the maximal available SHDSL bit rate of symmetrical wire pairs. The result of this prediction is the label of one of the bit rate groups. Two types of rule bases were created. One of them was constructed directly from the measured values. It consists of five six dimensional rules, in which triangular antecedent and consequent fuzzy sets are used. The input dimensions are the insertion loss values that can be measured at the 6 characteristic frequencies. Each rule can be unambiguously assigned to one of the output states. Using the measured and clustered data as teaching samples, another type of rule bases were created by bacterial evolutionary algorithm [4], resulting in a rule base with trapezoidal fuzzy sets and ten rules. This rule base has the same 6 dimensions: the 6 characteristic frequencies [5]. Examples of the rule antecedents from each type of rule base can be seen in Fig. 1. In the figure, the upper diagram belongs to the triangular rule base, and the lower one to the rule base constructed by bacterial evolutionary algorithm. The upper rule is obviously sparse. Even though seemingly there are no gaps between the trapezoidal fuzzy sets in the rule base made by the bacterial evolutionary algorithm, this rule base can be also considered as a sparse or incomplete one, because insertion loss values can be measured outside of the supports of all fuzzy sets, as well.

The above two rule bases were tested by the measurements of more than 60 wire pairs in operating access networks and there were no relevant differences between their respective results. In most of the cases, where all measured values belonged to insertion loss areas covered by antecedent sets, the predictions were successful. Only 13 lines out of 65 could be evaluated, and the predictions were correct in case of 12 lines form this 13.

Fig. 1 Examples of rule antecedents from our previous predicting methods [6]

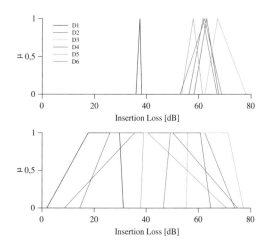

Fig. 2 Success rate of the
rule bases

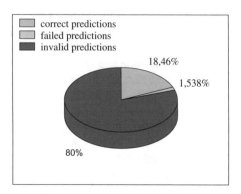

However correct the results of the successful predictions were, the high proportion (80%) of the lines (Fig. 2) where no results were produced due to the gaps in the rule base (no overlap with any antecedent), is not acceptable in practice. This phenomenon was caused by two reasons.

The reason for unsuccessful evaluations, where no valid results were produced, was the insufficient knowledge of the observed physical phenomenon. Obtaining only a limited amount of valid data during the measurement, only sparse rule bases could be constructed. Measurement results of wire pairs to be tested contained values also outside of the supports of the antecedent fuzzy sets, hence, in these cases there could be no valid conclusion calculated. This deficiency of the method is considered as a "horizontal sparseness".

The reason for valid, but incorrect prediction was investigated as well. This was caused by the drastically lowered dimensionality of the input space (values only at 6 points out of 150 were considered). Even though, generally, the insertion loss measured at the selected characteristic frequencies led to correct predictions, in several, infrequently occurring cases, measured values at these 6 frequency points could have such deviations which could bias the result of the prediction. Using only several variables from the whole input space can be considered as "vertical sparseness" of the model of the reality.

In the next section a new approach will be proposed, suitable for the elimination of both mentioned problems.

3   Methods for Handling the Vertical and Horizontal Sparsenesses

Vertical sparseness of the rule bases was derived from the partial usage of the possible input data. It was needed in order to decrease the dimensionality of the applied fuzzy inference system, however, a large amount of information of the measured insertion loss functions was wasted. Finding a method which keeps the simplicity of the fuzzy system and the information of the used insertion loss functions was needed. As wavelet transformation is efficient in reducing the size of any continuous or discrete functions down to a required level, it seemed to be successfully applicable in the problem.

Horizontal sparseness of the fuzzy system, namely the sparseness of the rule bases, can be handled by the techniques of fuzzy rule interpolation. Stabilized KH interpolation fits continuous and mathematically stable functions to all α-cuts of the membership functions in the rules, which can treat the observations in the gaps and out of the domains of the rules, too (in this way performing also extrapolation).

Basics of wavelet transformation and stabilized KH interpolation are briefly overviewed in the followings.

3.1 On Wavelet Analysis

In data processing in general wavelet theory [7, 8] has proved to be an extremely useful tool. The largest part of the methods using wavelets is the image [9] and video compression [10–12]. The first large-scale application was a fingerprint compression method [13], and also the compressors used in space equipments are mostly based on wavelet analysis [14], but one of the pioneering application was seismic signal analysis [15], which is a 1D data set similar to insertion loss values in many ways. Wavelet analysis from our point of view is similar to the discretized versions of Fourier analysis. In the next paragraphs their similarities and the differences will be summarized.

Both continuous wavelet transform and Fourier analysis [16] of a function provides data about the function's fine-scale and rough-scale behavior. Similarly, the discrete transforms turn the initial data vector or distribution into a set of coefficients corresponding to higher frequencies and lower frequencies. The well-known windowed Fourier transform of a function f can be summarized by the formula

$$\mathcal{F}_b\{f\}(\omega) = \int_{-\infty}^{\infty} w(x - b)\, f(x)\, e^{-i\omega x} dx, \tag{1}$$

with the window function w, which usually is either compactly supported or has very quickly decaying tails towards the positive and negative infinity. This provides a short sample of the function f to be transformed. However, if the window function is combined with the transforming exponential, a short wave

$$w_{b,\omega}(x) = e^{i\omega x}\, w(x - b),$$

is produced as a transforming function, and the windowed Fourier transform turns into

$$\mathcal{F}_b\{f\}(\omega) = \int_{-\infty}^{\infty} w_{b,\omega}(x)\, f(x) dx. \tag{2}$$

Wavelet transform is summarized as

$$\mathcal{W}_\psi\{f\}(ba) = |a|^{-1/2} \int_{-\infty}^{\infty} \psi_{b,a}(x) f(x) dx,$$

with the transforming function

$$\psi_{ba}(x) = |2|^{-a/2} \psi\left(\frac{x-b}{2^a}\right).$$

It can be seen from the above formula, that all the wavelets $\psi_{b,a}(x)$ are generated from one mother wavelet $\psi(x)$ by dilating and shifting, thus whereas in case of the windowed Fourier transform, the grid distance and the size of the transforming function remains the same at all resolutions, only its shape changes, for wavelet transforms, the shape of the functions remains the same and both the grid and the width of the function shrinks as the resolution (i.e., the frequency or spatial frequency) increases. This property is one of the driving forces behind the success of wavelet analysis, as usually the high frequency terms are localized in smaller spatial domains than the slowly varying parts, thus changing the window size with the frequency is usually very effective.

Wavelet analysis can be carried out by a series of filter pairs. There is a high-pass and a low-pass filter in all of the pairs, as it can be seen in Fig. 3, the high-pass ones (after a downsampling) giving the wavelet components and the low-pass ones being transformed further. The downsampling steps in each of the branches in Fig. 3 take only every second element of the result and neglect the points in between. In each of such steps the frequency limit of the low-pass distribution is at around the half of the highest frequency of the incoming distribution. It can be seen easily, that low pass outputs give a coarse-grained, or averaged behaviour of the distribution, whereas the wavelet terms provide the fine-scale details.

The total number of the elements in the resulting vectors c_i' and d_i' is almost the same as that of the original vector c_i, only a slight increase might arise due to the size of the filters. The size of the filters N_s can be different for the different wavelet types, however it is typically less than 20—in image processing usually less than

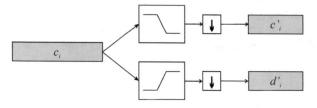

Fig. 3 One filter pair of the discrete wavelet transform. After the high pass and low pass convolutional filters and the downsamplings the transformed vectors c_i' and d_i' arise, their size is about half of the size of the original c_i

8, so this increment of N_s compared to the size of the data is usually negligible. Moreover, usually many of the vector components d'_i can be omitted as the fine details are localized to small fractions of the data, thus the resulting vectors are usually significantly (sometimes by one or two orders of magnitudes) smaller than the size of the original data, hence the data compression ability [9, 14, 17].

Mathematically, the wavelet transforms of a function can be written as

$$c_{ba} = \int_{-\infty}^{\infty} f(x)\phi_{ba}(x). \tag{3}$$

for the low-pass coefficients of resolution level 2^{-a}, and as

$$d_{ba} = \int_{-\infty}^{\infty} f(x)\psi_{ba}(x). \tag{4}$$

for the high-pass coefficients. The transforming function ϕ_{ba} is a so called scaling function, and ψ_{ba} a wavelet, as previously. From these coefficients the function f can be approximated at resolution level 2^{-A} as

$$f^A(x) = \sum_{-\infty}^{\infty} c_{b0}\phi_{b0}(x) + \sum_{a=0}^{A-1}\sum_{-\infty}^{\infty} g_{ba}\phi_{ba}(x), \tag{5}$$

One step of this synthesis procedure of the wavelet transform is summarized in Fig. 4, where the synthesis filters are after an upsampling step that introduces zeros between the elements of the incoming vectors. Usually, more of these steps are following each other with introducing finer and finer resolution details.

In data analysis—also in our case—the starting point is a sampled function and the end result is the lowest resolution level low pass vector and the high pass vectors. Our starting vector is a series of insertion loss values measured at consecutive frequency points, and the resulting vectors will give information about the large-scale behavior of the insertion loss vs. frequency function. In the following considerations

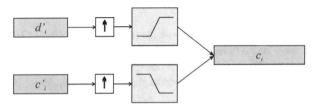

Fig. 4 Synthesis steps of the wavelet transform. The synthesis filters are plotted with yellow rectangles and the upsampling step with a rectangle including an upright arrow

Daubechies's [7] wavelet and scaling function sets are used with 2 and 4 nonzero filter coefficients. In case of the 2-element filter, the averaging process is without overlaps with the neighboring frequency domains, whereas, in case of the 4-element filter, the domains of the weighted averages overlap.

Transformations of the starting sampled insertion loss functions were carried out until only 10 and 5 vector elements remained. As the power spectral density of the transmission is larger in the lower frequency domain, we have merged the results of the two vectors so that the points would meet our previously selected characteristic frequency points.

3.2 Stabilized KH Rule Interpolation

In case of sparse rule bases, KH interpolation [18, 19] is a mathematically stable and widely applicable fuzzy rule interpolation method. Its improved version is the stabilized KH interpolation. In our work we used this improved technique in order to eliminate the problems originating from the sparseness of the rule bases.

The method is based on the distances between the examination vector and the antecedent sets of the rule base. The closures of the α-cuts of the interpolated resolution are given by

$$\inf\{B_\alpha^*\} = \frac{\sum_{i=1}^{2n} \left(\frac{1}{d_{\alpha L}(A^*, A_i)}\right)^k \inf\{B_{i\alpha}\}}{\sum_{i=1}^{2n} \left(\frac{1}{d_{\alpha L}(A^*, A_i)}\right)^k} \tag{6}$$

and

$$\sup\{B_\alpha^*\} = \frac{\sum_{i=1}^{2n} \left(\frac{1}{d_{\alpha U}(A^*, A_i)}\right)^k \sup\{B_{i\alpha}\}}{\sum_{i=1}^{2n} \left(\frac{1}{d_{\alpha U}(A^*, A_i)}\right)^k}, \tag{7}$$

where i denotes the number of the rules, k the number of the dimensions (variables), A^* the observation, A_i the antecedent sets in rule i, $d_{\alpha L}(A^*, A_i)$ and $d_{\alpha U}(A^*, A_i)$ the lower and upper bounds of the distance between the α-cuts of observation and the antecedents, and B^* stands for the corresponding fuzzy conclusion [20]. In practice it is efficient to calculate the values of B_α^* for $\alpha = 0$ and 1.

4 A New Performance Prediction Method Based on the Combination of Wavelet Transformation and Stabilized KH Rule Interpolation

In order to avoid the problems reviewed in Sect. 1, the techniques of Sects. 3.1 and 3.2 were used.

First, the wavelet transformed version of the insertion loss values used in rule base construction were calculated. Daubechies-2 (Haar) [21] and Daubechies-4 wavelets were used and the transformations were performed down to 5 points resolution as described in the previous section. Figure 5 shows the original and the Haar wavelet transformed insertion loss values as an example. As a matter of course, wavelet transformation results in discrete values, however, to make the corresponding points visible, they are graphically linked in the figure.

New rule bases were created by the Wavelet transformed insertion loss series. In this pattern, the rule base based on Daubechies wavelets did not give better results than the old one without any wavelet transformation, moreover, several additional errors were detected. On the contrary, in case of the rule base made by Haar wavelets, accurate results were gained for each of the 13 line that produce valid results and one further line could be assessed, too, as it can be seen in Fig. 6.

In order to evaluate those lines that were previously not to be assessed, this new (Haar wavelets based) rule base was applied together with the stabilized KH rule interpolation. The 65 test lines were re-processed, thus the predictions became feasible in the case of all lines. The predictions for the 13 wire pairs which were correctly

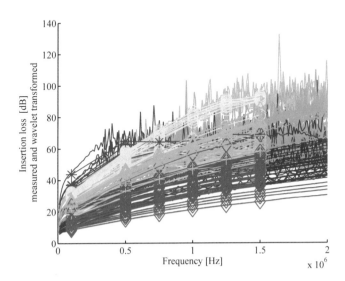

Fig. 5 Insertion loss values and the corresponding wavelet transforms. Different performance classes are indicated by different colors

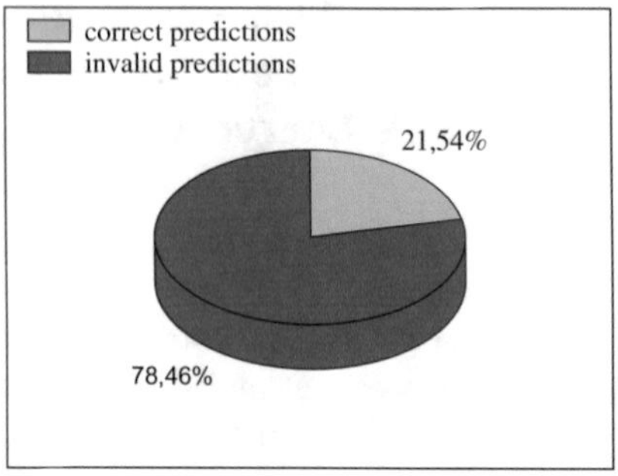

Fig. 6 Efficiency of the rule base based on Haar wavelets

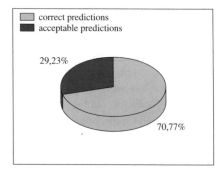

Fig. 7 Efficiency of the Haar wavelets based rule base supplemented with the stabilized KH rule interpolation

evaluated previously remained valid, moreover, results of the predictions of 33 from the other 52 were correct, and 19 acceptable (in this contribution, results with a deviation of −1 from the correct values are considered as acceptable ones, all the others as incorrect) and there were no incorrect results see Fig. 7.

The simplified "algorithm" of the construction of the predicting system is as follows.

- Collection of insertion loss and bit rate data of wire pairs.
- Dividing the whole bit rate domain into groups (the more the number of the measured lines, the finer is the possible resolution) and clustering the measured values into these groups.
- Generation of several discrete values (6 in this case, however, other resolutions are examined by our ongoing investigations) from measured insertion loss functions by wavelet transformation (Haar wavelets are now recommended, though investigating other types of wavelets with other resolution levels are being in progress).
- Construction of fuzzy rule bases by clustered and wavelet transformed values.
- Wavelet transformation of the insertion loss function of the wire pair to be predicted.
- Prediction making by stabilized KH interpolation (can be made even if the input values can be found within the areas covered by antecedent fuzzy sets).

5 Conclusions

A novel performance prediction method based on interpolated fuzzy inference for telecommunications transmission lines and wavelet transformation of the values of the physical parameters influencing the performance was presented. The combination of the fuzzy rule interpolation and wavelet transformation was proposed in this paper in the first time. Wavelet transform was used for generating a coarse-grained view of the measured data, whereas the interpolation was applied for treating the sparseness of the rule bases. The method performed very well for the model system of the SHDSL connections, 52 predictions from 65 test cases were correct, and the other 19 were acceptable.

References

1. ITU.: Single-pair high-speed digital subscriber line (SHDSL) transceivers, Technical Recommendation G.991.2, ITU Std., Dec 2003
2. ITU.: Single-pair high-speed digital subscriber line (SHDSL) transceivers, amendment 2, Technical Recommendation G.991.2 amendment 2, ITU Std., Feb 2005
3. Lilik, F., Nagy, Sz., Kóczy, L.T.: Wavelet Based Fuzzy Rule Bases in Pre-Qualification of Access Networks Wire Pairs. IEEE Africon 2015, Addis Ababa, Ethiopia, 14–17 Sept 2015 paper P-52, 5 p
4. Balázs, K., Kóczy, L.T.: Constructing dense, sparse and hierarchical fuzzy systems by applying evolutionary optimization techniques. Appl. Comput. Math. **11**(1), 81–101 (2012)
5. Lilik, F., Botzheim, J.: Fuzzy based prequalification methods for EoSHDSL technology. Acta Tech. Jauriensis Ser. Intelligentia Computatorica **4**(1), 135–145 (2011)
6. Lilik, F., Kóczy, L.T.: The determination of the bitrate on twisted pairs by Mamdani inference method, issues and challenges of intelligent system and computational intelligence. Stud. Comput. Intell. **530**, 59–74 (2014). https://doi.org/10.1007/978-3-319-03206-1_5
7. Daubechies, I.: Ten Lectures on Wavelets. CBMS-NSF Regional Conference Series in Applied Mathematics 61. SIAM, Philadelphia (1992)
8. Chui, C.K.: An Introduction to Wavelets. Academic Press, San Diego (1992)
9. Christopoulos, Ch., Skodras, A., Ebrahimi, T.: The JPEG2000 still image coding system: an overview. IEEE Trans. Consum. Electron. **46**, 1103–1127 (2000). https://doi.org/10.1109/30.920468
10. MJ2 format standard, ISO/IEC 15444-3:2002/Amd 2:2003
11. Martucci, S.A., Sodagar, I., Chiang, T., Zhang, Y.-Q.: A zerotree wavelet video coder. IEEE Trans. Circ. Sys. Video Technol. **7**, 109–118 (1997)
12. Chai, B.-B., Vass, J., Zhuang, X.: Significance-linked connected component analysis for wavelet image coding. IEEE Trans. Image Proc. **8**, 774–784 (1999)
13. Montoya Zegarra, J.A., Leiteb, N.J., da Silva, Torres R.: Wavelet-based fingerprint image retrieval. J. Comput. Appl. Math. **227**, 297–307 (2008). https://doi.org/10.1016/j.cam.2008.03.017
14. Kiely, A., Klimesh, M.: The ICER Progressive Wavelet Image Compressor. IPN Progress Report 42-155, 15 Nov 2003. http://ipnpr.jpl.nasa.gov/tmo/progressreport/42-155/155J.pdf
15. Grossmann, A., Morlet, J.: Decomposition of hardy functions into square integrable wavelets of constant shape. SIAM J. Math. Anal. **15**, 723–736 (1984)
16. Fourier, J.B.J.: Theorie Analitique de la Chaleur. Didot, Paris (1822)

17. Nagy, Sz., Pipek, J.: On an economic prediction of the finer resolution level wavelet coefficients in electron structure calculations. Phys. Chem. Chem. Phys. **17**, 31558–31565 (2015). https://doi.org/10.1039/C5CP01214G

18. Kóczy, L.T., Hirota, K.: Approximate reasoning by linear rule interpolation and general approximation. Int. J. Approximate Reasoning **9**, 197–225 (1993). https://doi.org/10.1016/0888-613X(93)90010-B

19. Kóczy, L.T., Hirota, K.: Interpolative reasoning with insufficient evidence in sparse fuzzy rule bases. Inf. Sci. **71**, 169–201 (1993). https://doi.org/10.1016/0020-0255(93)90070-3

20. Tikk, D., Joó, I., Kóczy, L.T., Várlaki, P., Moser, B., Gedeon, T.D.: Stability of interpolative fuzzy KH-controllers. Fuzzy Sets Syst. **125**, 105–119 (2002). https://doi.org/10.1016/S0165-0114(00)00104-4

21. Haar, A.: Zur Theorie der Orthogonalen Funktionensysteme. Math. Ann. **69**, 331–371 (1910)

Abductive Reasoning on Molecular Interaction Maps

Jean-Marc Alliot, Robert Demolombe, Luis Fariñas del Cerro, Martín Diéguez and Naji Obeid

Abstract Metabolic networks, formed by a series of metabolic pathways, are made of intra-cellular and extracellular reactions that determine the biochemical properties of a cell, and by a set of interactions that guide and regulate the activity of these reactions. Cancer, for example, can sometimes appear in a cell as a result of some pathology in a metabolic pathway. Most of these pathways are formed by an intricate and complex network of chain reactions, and are often represented in *Molecular Interaction Maps* (MIM), a graphical, human readable form of the cell cycle checkpoint pathways. In this paper, we present a logic, called Molecular Interaction Logic, which semantically characterizes MIMs and, moreover, allows us to apply deductive and abductive reasoning on MIMs in order to find inconsistencies, answer queries and infer important properties about those networks.

1 Introduction

Metabolic networks, formed by series of metabolic pathways, are made of intra-cellular and extracellular reactions that determine the biochemical properties of a cell, and by a set of interactions that guide and regulate the activity of these reactions. These reactions, which can be positive (production of a new protein) or negative (inhibition of a protein in the cell), are at the center of a cell's existence and they can

This research was partially supported by the French Spanish Laboratory for Advanced Studies in Information, Representation and Processing (LEA-IREP). Martín Diéguez was supported by the Centre international de mathématiques et d'informatique (contract ANR-11-LABX-0040-CIMI).

J.-M. Alliot · R. Demolombe · L. Fariñas del Cerro · M. Diéguez (✉) · N. Obeid
University of Toulouse, CNRS, IRIT, Toulouse, France
e-mail: Martin.Dieguez@irit.fr

© Springer International Publishing AG 2018
L. T. Kóczy and J. Medina (eds.), *Interactions Between Computational Intelligence and Mathematics*, Studies in Computational Intelligence 758,
https://doi.org/10.1007/978-3-319-74681-4_4

be modulated by other proteins, which can either enable these reactions or, on the opposite, inhibit them.

Medical and pharmaceutical researches [15, 19] showed that the break of the double strand of DNA sometimes appears in a cell as a result of some pathology in a metabolic pathway, and double strand break (*dsb*) is a major cause of cancer.

These pathways are used to investigate the molecular determinants of tumor response in cancers. The molecular parameters include the cell cycle checkpoint, DNA repair and apoptosis[1] pathways [15, 19, 21, 25, 26]. When DNA damage occurs, cell cycle checkpoints are activated and can rapidly kill the cell by apoptosis or arrest the cell cycle progression to allow DNA repair before cellular reproduction or division (see, for instance, the *atm-chk2* and *atr-chk2* pathways in [26]).

Most of these pathways are formed by an intricate and complex network of chain reactions, and are often represented in *Molecular Interaction Maps* (MIM), a human readable form of the cell cycle checkpoint pathways. MIMs become increasingly larger and their density is constantly enriched with new information (references, date, authors, etc.). Although essential for knowledge capitalization and formalization, MIMs are difficult to use:

- Reading is complex due of the very large number of elements, and reasoning about the map is even more difficult.
- Using a map to communicate goals is only partially suitable because the representation formalism requires expertise.
- Maps can contain implicit knowledge, that is taken for granted by one expert, but is missed by another one.
- Maps can be inconsistent, and these inconsistencies are difficult to detect just by looking at the map itself.

These problems have been faced from the point of view of nonmonotonic reasoning, specially Answer Set Programming (ASP) [13, 14] or action languages [2], taking advantage of non-monotonictiy and the efficient ASP solvers that are available. Our approach is based on a reduction to classical logic which allows applying classical reasoning on such kind of networks.

In this paper we present a method that automatically transforms a MIM into a set of logical formulas, taking as input the XML files generated by a graphical editor for biological pathways such as pathvisio [31]. Along this paper we will use, as examples, subgraphs of the pathway of Fig. 1, which represents the modelling of the *atm-chk2* pathway leading to apoptosis.

The rest of this paper is organized as follows: Sect. 2 introduces the concept of Molecular Interaction Maps and how they can be translated into a set of logical formulas. Section 3 describes Molecular Interaction Logic, a logic which is capable of describing and reasoning about general pathways. Section 4 investigates the application of deductive as well as abductive reasoning on MIMs. Section 5 presents the tools used and/or implemented and Sect. 6 investigates the future lines of work.

---

[1] Apoptosis is the process of programmed cell death.

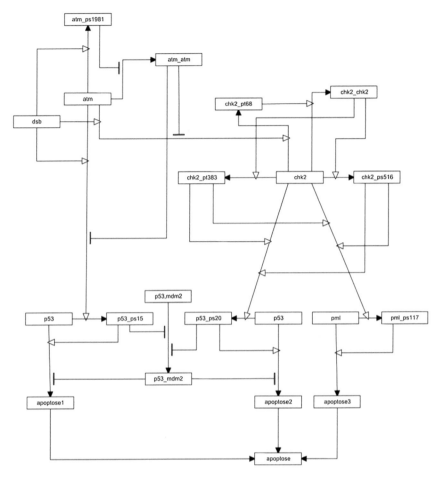

Fig. 1 $atm - chk2$ pathway

2   Molecular Interaction Maps

A Molecular Interaction Map [20] (MIM) is a diagram convention which represents the interaction networks of multi-protein complexes, protein modifications and
enzymes that are substrates of other enzymes. Although interactions between elements of a MIM can be complex, they can be represented using only three basic
connectors: production (→), activation (⇢) and inhibition (⊣). Figure 1 presents
the atm-$chk2$ pathway, using only the aforementioned connectors.

A production relation means that a new substance is created as a result of a reaction
on several primary components. For instance, the protein atm can be dimerized
to become the atm_atm protein or phosphorylated at serine 1981 resulting in the
production of atm_ps1981. These reactions can be triggered or blocked by other

proteins or conditions. For example, in Fig. 1, atm_ps1981 blocks the dimerization of atm into atm_atm, while the double strand break (dsb) of DNA triggers the production of atm_ps1981 by atm.

These interactions can be "stacked": for example, protein $p53$ can be phosphory-lated at serine 15 to become $p53\_ps15$ (see Fig. 1). This reaction is triggered by atm, but the triggering itself has to be activated by dsb and can be blocked by atm_atm. Thus, the two main actions (production of a protein or inhibition of a protein) can be triggered or blocked by a stack of preconditions.

2.1 Translation of MIMs into Formulas

Our first goal is to translate any MIM into a set of logical expressions in order to perform several automated reasoning tasks such as deduction or abduction. First, focusing on the diagram of Fig. 2 (which corresponds to a sub-diagram of Fig. 1) will help getting an intuitive idea of how translation is performed.

Here *apoptosis* arises when protein $p53$ is phosphorylated at serine 20 or 15 (instances $p53\_ps20$ and $p53\_ps20$ respectively). However, *apoptosis* would not happen if the dimer $p53\_mdm2$ is present. Thus the context would be *if $p53$ and either $p53\_ps20$ or $p53\_ps15$ are present and $p53\_mdm2$ is absent then apoptosis is produced* (this example should of course be completed with the rules for producing the rest of objects in the diagram).

The general form of production relations is displayed in Fig. 3.

Each arrow can be either an activation or an inhibition of the relation it applies to, and these activations/inhibitions can be stacked on any number of levels. The above examples give the idea behind the translation: it is a recursive process starting from the production relation and climbing up the tree. In order to formally describe these graphs, we define below the concepts of *pathway context* and *pathway formula*.

Definition 1 (*Pathway context*) Given a set of entities, a pathway context is formed by expressions defined by the following grammar:

$$\alpha ::= \langle \alpha P \twoheadrightarrow, \alpha Q \dashv \rangle | \langle P \twoheadrightarrow, Q \dashv \rangle$$

Fig. 2 Apoptosis by $p53\_ps20$ and $p53\_ps15$ mediation

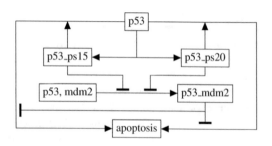

Fig. 3 The general form of a
basic production

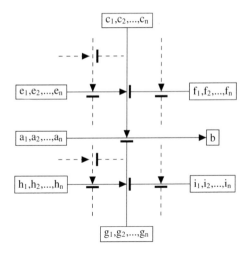

where P and Q are sets (possibly empty) of propositional variables representing the
conditions of activation ($\rightarrow$) or inhibition ($\dashv$) of the reaction. The first part of the
pair is the activation context, the second part is the inhibition context. One, or both
sets can be empty. ∎

For example, the $p53 \rightarrow apoptosis$ reaction of Fig. 2 would lead to the following
two pathway contexts:

$$\langle p53\_ps20 \rightarrow, p53\_mdm2 \dashv \rangle \tag{1}$$
$$\langle p53\_ps15 \rightarrow, p53\_mdm2 \dashv \rangle \tag{2}$$

Definition 2 (*Causal pathway formulas*) A *causal pathway formula* is defined by
the following grammar:

$$F ::= [\alpha](p_1 \wedge \cdots \wedge p_n \rightarrow \textbf{Pr } q) \mid$$
$$[\alpha](p_1 \wedge \cdots \wedge p_n \rightarrow \textbf{In } q) \mid$$
$$F \wedge F$$

where α is a pathway context, $p_1, \ldots, p_n, q$ are propositional variables while **Pr**
and **In** are modal concepts that qualify the process of activation or inhibition of
proteins. ∎

Applied to the example of Fig. 2, the causal pathway formula associated with the
production rule $p53 \rightarrow apoptosis$ is

$$[(1)](p53 \rightarrow \textbf{Pr} \ apoptosis) \wedge$$
$$[(2)](p53 \rightarrow \textbf{Pr} \ apoptosis) . \tag{3}$$

Observation 1 *Each MIM can now be represented as a causal pathway formula.* ∎

3 Semantics for Causal Pathway Formulas

In this section the semantics of causal pathway formulas is formally introduced. The resulting logic, denoted by MIL (Molecular Interaction Logic), extends a previous work [4, 5] where the MIMs were formalized via first order logic with equality, in which the pathway contexts were limited to one level of depth. From now on, p means *protein p is present* and $\neg p$ means *protein p is absent*. Before going into details, we first provide a method to translate a pathway context into a classical Boolean expression, which will be used in the definition of the satisfaction relation.

Definition 3 (*Activation and inhibition expressions*) Given a pathway context $\alpha = \langle \alpha' P \rightarrow, \beta' Q \dashv \rangle$, the *activation* and the *inhibition expressions* associated with the context α (denoted by $A(\alpha)$ and $I(\alpha)$) are defined recursively as:

$$A(\alpha) = \bigwedge_{p \in P} p \wedge A(\alpha') \wedge (\bigvee_{q \in Q} \neg q \vee I(\beta'))$$

$$I(\alpha) = \bigvee_{p \in P} \neg p \vee I(\alpha') \vee (\bigwedge_{q \in Q} q \wedge A(\beta'))$$

The above expressions define the general forms of $A(\alpha)$ and $I(\alpha)$. If one part of the context α is empty, then the corresponding part is of course absent in $A(\alpha)$ and $I(\alpha)$. ∎

Following such definition, formulas associated with (1) are:

$$A((1)) = p53\_ps20 \wedge \neg p53\_mdm2$$
$$I((1)) = \neg p53\_ps20 \vee p53\_mdm2$$

Definition 4 (*MIL interpretation*) A MIL interpretation consists of a pair (V_1, V_2) of classical evaluations i.e. $V : \mathcal{P} \rightarrow \{True, False\}$ where $\mathcal{P}$ is the set of propositional variables. ∎

The intuitive meaning behind these two evaluations correspond for V_1 to the protein present or absent, and for V_2 to the state of the protein resulting from the chemical reactions in the cell.[2]

Definition 5 (*Satisfaction relation*) Given a MIL interpretation (V_1, V_2), the satisfaction relation is defined as:

(1) $(V_1, V_2) \vDash p$ iff $V_1(p) = True$ for $p \in \mathcal{P}$;
(2) $\wedge, \vee$ and $\rightarrow$ are satisfied as in classical logic;
(3) $(V_1, V_2) \vDash \mathbf{Pr}\ p$ iff $V_1(p) = V_2(p) = True$, with $p \in \mathcal{P}$;
(4) $(V_1, V_2) \vDash \mathbf{In}\ p$ iff $V_1(p) = V_2(p) = False$, with $p \in \mathcal{P}$;
(5) $(V_1, V_2) \vDash [\alpha]F$ iff $(V_1, V_2) \nvDash A(\alpha)$ or $(V_1, V_2) \vDash F$, for any pathway formula $[\alpha]F$.

■

While the first four satisfaction relations are simple to understand, the fifth one is a little bit trickier according to Definition 3; its meaning is that if the conditions of activation of α are satisfied, then the reaction represented by F holds. As usual, a formula F is satisfiable if there is a model (V_1, V_2) such that $(V_1, V_2) \vDash F$.

Observation 2 *MIL can be characterized by the axioms of classical logic, plus the axioms:*

1. $[\alpha]F \leftrightarrow (A(\alpha) \rightarrow F)$
2. $\mathbf{Pr}\ p \rightarrow p$, *if p is produced then p is present*
3. $\mathbf{In}\ p \rightarrow \neg p$, *if p is inhibited then p is absent*

■

As a result of MIL semantics, the causal pathway formula (3) is logically equivalent to the conjunction of the following implications:

$$(p53 \wedge p53\_20 \wedge \neg p53\_mdm2) \rightarrow \mathbf{Pr}\ apoptosis \qquad (4)$$
$$(p53 \wedge p53\_20 \wedge \neg p53\_mdm2) \rightarrow \mathbf{Pr}\ apoptosis \qquad (5)$$

Observation 3 *Any MIM can be transformed into a causal pathway formula, and every causal pathway formula is equivalent to a boolean composition of:*

– *propositional variables or their negation*
– *propositional variables qualified by* **Pr** *or* **In** *or their negation*

■

Axioms (2) and (3) of Observation 2 have as consequence:

---

[2]If the semantics of the modal logic S5 is restricted to have at most two worlds then a strong normal form in which conjunctions and disjunctions are not in the scope of a modal operator can be found for this new logic [11]: the pathway causal formulas of MIL verify this condition.

Observation 4 *Given a MIM formula F, adding* **Pr** $p \rightarrow p$ *and* **In** $p \rightarrow \neg p$ *for each propositional variable p in F, enables us to embbed MIL into classical logic.* ∎

The notions of completion [3] and production axioms, which are both important and implicit in MIMs, are presented first.

Definition 6 (*Completion Axiom from* [3]) Let *p* be an object that can be produced through different pathways:

$$C_1 \rightarrow \textbf{Pr} \ p$$
$$\cdots$$
$$C_n \rightarrow \textbf{Pr} \ p$$

Every C_i formula represents one pathway leading to *p*. The completion axiom for **Pr** p is the implication **Pr** $p \rightarrow (C_1 \vee \cdots \vee C_n)$. ∎

This axiom means that **Pr** p is a notion local to the current map and, therefore, *p* has to be produced by at least one of the C_i possible pathways. For instance, pathway rules related to *apoptosis* in Fig. 2 would lead to the following completion axiom:

Pr $apoptosis \rightarrow ((p53 \wedge p53\_ps20 \wedge p53\_mdm2) \vee (p53 \wedge p53\_ps15 \wedge p53\_mdm2))$,

which corresponds to all the pathways producing apoptosis in Fig. 2.

Before presenting the second axiom, the term *endogenous entity* has to be introduced. Endogenous entities are the ones which have to be produced in the map in order to be present.

It is worth to mention that the concept of endogenous entity is a biological notion and is moreover local to a given MIM; it is up to the biologists to mark in an interaction map which objects they consider as endogenous. This is an information which is not currently present in MIMs, as it is implicit knowledge for biologists.[3]

Definition 7 (*Production axiom*) Given an *endogenous* entity *p* the *production axiom* associated with *p* is defined by $p \rightarrow \textbf{Pr} \ p$. ∎

For instance, the logical representation of the diagram of Fig. 2 would require the addition of such axioms for *apoptosis*, *p53_mdm2*, *p53_ps20* and *p53_ps15*, the endogenous entities occurring in this map.

---

[3]In a broadly simplified way, endogenous entities can be considered as "internal variables" or "output variables" of the model, while exogenous entities are "command variables".

4 Deductive and Abductive Reasoning on MIMs

In this section, deductive and abductive reasoning are used on MIMs in order to find inconsistencies in the representations and to answer questions about production or inhibition of proteins in a MIM.

4.1 Deductive Reasoning

Figure 4 represents the dimerization of *atm* into *atm_atm*. When translated into its logical representation, a SAT-checker finds that this representation is inconsistent. Why so? When *atm* and not *dsb* are present then *atm_atm* is produced. However, with *dsb*, *dsb_atm_atm* enables the phosphorylation of *atm* into *atm_ps*1981 which then blocks *atm_atm*, which is inconsistent with the fact that *atm_atm* is necessary to produce *dsb_atm_atm*.

This inconsistency arises because biologists (at least some of them) have some implicit temporal knowledge about the way these reactions take place, but they do not represent this knowledge explicitly in MIMs. This shows that if MIMs are to become a consistent medium for representing proteins interaction, they have to be enriched with temporal informations.

4.2 Abductive Reasoning

In this section abductive reasoning is used to answer queries formulated on MIMs. The diagram of Fig. 2 will be used as an example. Its logical encoding consists of the conjunction of the implications given below:

Fig. 4 The *dsb_atm_atm* loop

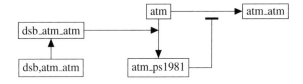

$$p53 \wedge p53\_ps20 \wedge \neg p53\_mdm2 \rightarrow \textbf{Pr } apoptosis \tag{6}$$

$$p53 \wedge p53\_ps15 \wedge \neg p53\_mdm2 \rightarrow \textbf{Pr } apoptosis \tag{7}$$

$$mdm2 \wedge \neg p53\_ps20 \wedge p53 \wedge$$
$$\neg p53\_ps15 \rightarrow \textbf{Pr } p53\_mdm2 \tag{8}$$

$$p53 \rightarrow \textbf{Pr } p53\_ps20 \tag{9}$$

$$p53 \rightarrow \textbf{Pr } p53\_ps15. \tag{10}$$

These formulas correspond to 5 different pathways, two describing the production of *apoptosis*, two defining the phosphorylation of the protein *p53* into *p53_ps20* and *p53_ps15* respectively and, finally, one representing the production of the dimer *p53_mdm2* from *mdm2* and *p53*. The corresponding completion axioms are:

$$\textbf{Pr } apoptosis \rightarrow (p53 \wedge p53\_ps20 \wedge \neg p53\_mdm2) \vee$$
$$(p53 \wedge p53\_ps15 \wedge \neg p53\_mdm2) \tag{11}$$

$$\textbf{Pr } p53\_mdm2 \rightarrow mdm2 \wedge \neg p53\_ps20 \wedge p53 \wedge \neg p53\_ps15 \tag{12}$$

$$\textbf{Pr } p53\_ps20 \rightarrow p53 \tag{13}$$

$$\textbf{Pr } p53\_ps15 \rightarrow p53. \tag{14}$$

The production axioms must now be added for all endogenous proteins: $p53\_ps15 \rightarrow \textbf{Pr } p53\_ps15, p53\_ps20 \rightarrow \textbf{Pr } p53\_ps20, p53\_mdm2 \rightarrow \textbf{Pr } p53\_mdm2$, and $apoptosis \rightarrow \textbf{Pr } apoptosis$.

The database is now complete and the system can be queried using abductive methods (see Sect. 5); for example, the question ?*apoptosis* gives the answer: $apoptosis \vee p53\_ps20 \vee p53\_ps15 \vee p53$. These results are coherent from a biological point of view. If *p_53* is present, then *apoptosis* occurs, but the same conclusion can be obtained from either *p53_ps15* or *p53_ps20*, since any of them is present if *p53* is.

The diagram of Fig. 2 is now modified by adding a new protein, *chk2*, which is required to activate the production of *p53_ps20*.

This new representation (Fig. 5) implies that rule (9) becomes $p53 \wedge chk2 \rightarrow \textbf{Pr } p53\_ps20$ and the corresponding completion axiom (13) becomes: $\textbf{Pr } p53\_ps20 \rightarrow p53 \wedge chk2$.

When applying abductive reasoning on the new diagram, the answer is pretty unexpected:

$$(apoptosis) \vee (\neg p53\_mdm2 \wedge p53\_ps15) \vee$$
$$(p53 \wedge \neg p53\_mdm2) \vee (p53\_ps15 \wedge \neg mdm2) \vee$$
$$(p53\_ps15 \wedge chk2) \vee (p53 \wedge \neg mdm2) \vee$$
$$(p53 \wedge chk2) \vee (p53\_ps20).$$

Fig. 5 Modification of the
diagram of Fig. 2

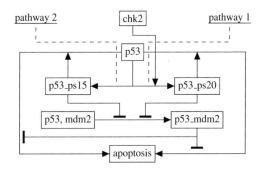

Now $p53$ alone is not enough for *apoptosis* anymore; ($p53$ and $chk2$), or ($p53$ and not $mdm2$) are needed for *apoptosis*. It is not immediately clear why the simple introduction of $chk2$ in the pathway labelled as pathway 2 changes the answer. This comes from the fact that, in this diagram, $p53\_ps15$ does not block the production of $p53\_mdm2$ explicitly. Therefore, if both $p53$ and $mdm2$ are present but $chk2$ is not, then $p53\_mdm2$ is produced in pathway 2 blocking *apoptosis* in pathway 1.

As a last, complicated test, the complete diagram in Fig. 1 which represents the whole $atm - chk2$ pathway was used. The logical representation of this diagram requires around 80 clauses. The abductive query (?*apoptosis*) was performed to find all conditions on *exogenous* variables that lead to *apoptosis*. The answer $((p53 \wedge dsb \wedge atm) \vee (pml \wedge chk2 \wedge dsb \wedge atm))$ are both correct and a little bit surprising. They clearly correspond to the paths leading to *apoptosis*1 and *apoptosis*3. The reason why the conditions leading to *apoptosis*2 do not appear becomes clear when solving ?*apoptosis*2 alone. The answer to this question $((p53 \wedge chk2 \wedge dsb \wedge atm))$ is subsumed by the one leading to *apoptosis*1.

Abductive queries give also the states of all the endogenous variables in the graphs which lead to apoptosis. They can be an invaluable tool for biologists on complex MIMs, where establishing such results by hand is both tedious and error-prone.

5 Implementation

All the examples presented here were written using *Pathvisio*, a public-domain MIM editing software.

A parser was developed to automatically translate the MIM from the XML generated by Pathvisio, add the formulas generated by the completion axioms and transform the formulas into Conjunctive Normal Form.

As demonstrated above, using Observation 4 we transform each set of clauses of our language in a set of classical clauses, allowing the use of both deduction and abduction algorithms for classical logic. Regarding deduction and SAT-checking, several efficient tools are available, such as the *minisat* solver [9].

However, regarding abductive reasoning, finding efficient systems able to solve quickly complex propositional problems is much more difficult. Different algorithms (see the survey [8] to get an overview of the state of the art) and systems [1, 24, 27] are present in literature. All those tools are implemented in pseudo-interpreted or interpreted languages such as Prolog, and they are slow when computing the whole set of prime implicates. The best solver found was SOLAR [23], which is a general purpose solver for abduction on first order theories with equality. When applied to our case, SOLAR could not take advantage of the fact that our theories are always propositional and, due to the complexity of general MIM's provided by the biologist, we developed a solver for computing prime implicates, which is based on the algorithm presented by Kean, Tsiknis [18] and Jackson [17].

This solver has been designed with only efficiency in mind. It uses bit representations for clauses and performs subsumptions and resolutions using machine instructions such as *and*, *or* and *xor*, with a parallel implementation using shared memory. Using this solver, the same example of Fig. 1 is solved using an ordinary laptop PC, with 185162519 subsumptions and 1519820 resolutions computed in 0.5 s. The detailed results of this work will be presented in another paper.

6 Conclusions and Future Work

We have presented a method to automatically translate MIMs into logical formulas: inconsistencies and missing knowledge in existing MIMs can be found, and abductive problems on quite complex MIMs can be automatically solved. There remain, however, different kind of problems:

- Currently the parser is only able to translate a subset of MIM relations; while this subset is enough to express all relations, many "shortcuts" relations in real MIMs have to be implemented to have an operational tool.
- While our implementation is able to solve abductive queries for an already quite complex MIM in a very short amount of time, it remains to be seen how it will behave on more realistic and thus much larger MIMs.
- MIMs have to be extended to represent reaction times, which are currently implicit and our logic will thus have to be extended to express reaction times.
- In order to reduce the research in a MIM, we plan to enrich the language with concepts like "aboutness" able to qualify, for example, proteins. It should allow us to isolate the subgraph of a given MIM, regarding the qualified proteins.

References

1. Alberti, M., Gavanelli, M., Lamma, E., Mello, P., Torroni, P.: The SCIFF abductive proof-procedure. In: AI*IA'05, p. 135147 (2005)
2. Baral, C., Chancellor, K., Tran, N., Tran, N., Joy, A., Berens, M.: A knowledge based approach for representing and reasoning about signaling networks. Bioinformatics **20** (suppl 1), i15–i22 (2004)

3. Clark, K.L.: Negation as failure. In: Logic and Databases, pp. 293–322. Plenum Press (1978)
4. Demolombe, R., Fariñas del Cerro, L., Obeid, N.: Logical model for molecular interactions maps. In: Logical Modeling of Biological Systems, pp. 93–123. Wiley, New York (2014)
5. Demolombe, R., Fariñas del Cerro, L., Obeid, N.: Translation of first order formulas into ground formulas via a completion theory. J. Appl. Logic (), to appear
6. Demolombe, R., Fariñas del Cerro, L.: An inference rule for hypothesis generation. In: IJCAI'91 (1991)
7. Demolombe, R., Fariñas del Cerro, L.: Information about a given entity: from semantics towards automated deduction. J. Logic Comput. **20**(6), 1231–1250 (2010)
8. Denecker, M., Kakas, A.: Abduction in logic programming. In: Computational Logic: Logic Programming and Beyond, pp. 402–436 (2002)
9. Een, N., Srensson, N.: An extensible sat-solver. In: SAT'03, pp. 502–518 (2003)
10. Erwig, M., Walkingshaw, E.: Causal reasoning with neuron diagrams. In: VLHCC '10, pp. 101–108 (2010)
11. Fariñas del Cerro, L., Herzig, A.: Contingency-based equilibrium logic. In: LPNMR'11, pp. 223–228 (2011)
12. Fariñas del Cerro, L., Inoue, K. (eds.): Logical Modeling of Biological Systems. Wiley, New York (2014)
13. Gebser, M., Guziolowski, C., Ivanchev, M., Schaub, T., Siegel, A., Thiele, S., Veber, P.: Repair and prediction (under inconsistency) in large biological networks with answer set programming. In: KR'10 (2010)
14. Gebser, M., Schaub, T., Thiele, S., Veber, P.: Detecting inconsistencies in large biological networks with answer set programming. Theory Pract. Logic Program. **11**(2–3), 323–360 (2011)
15. Glorian, V., Maillot, G., Poles, S., Iacovoni, J.S., Favre, G., Vagner, S.: Hur-dependent loading of mirna risc to the mrna encoding the ras-related small gtpase rhob controls its translation during uv-induced apoptosis. Cell Death Differ. **18**(11), 1692–1701 (2011)
16. Inoue, K.: Linear resolution for consequence finding. Artif. Intell. **56**(2–3), 301–353 (1992)
17. Jackson, P.: Computing prime implicates incrementally. In: CADE'92, pp. 253–267 (1992)
18. Kean, A., Tsiknis, G.: An incremental method for generating prime implicants/implicates. J. Symbolic Comput. **9**, 185–206 (1990)
19. Kohn, K.W., Pommier, Y.: Molecular interaction map of the p53 and mdm2 logic elements, which control the off-on swith of p53 response to dna damage. Biochem. Biophys. Res. Commun. **331**(3), 816–27 (2005)
20. Kohn, K.W., Aladjem, M.I., Weinstein, J.N., Pommier, Y.: Molecular interaction maps of bioregulatory networks: a general rubric for systems biology. Mol. Biol. Cell. **17**(1), 1–13 (2006)
21. Lee, W., Kim, D., Lee, M., Choi, K.: Identification of proteins interacting with the catalytic subunit of pp2a by proteomics. Proteomics **7**(2), 206–214 (2007)
22. Muggleton, S., Bryant, C.H.: Theory completion using inverse entailment. In: ILP'00, pp. 130–146 (2000)
23. Nabeshima, H., Iwanuma, K., Inoue, K., Ray, O.: SOLAR: an automated deduction system for consequence finding. AI Commun. **23**(2–3), 183–203 (2010)
24. Nuffelen, B.V.: A-system: problem solving through abduction. BNAIC01 Sponsors 1, 591–596 (2001)
25. Pei, H., Zhang, L., Luo, K., Qin, Y., Chesi, M., Fei, F., Bergsagel, P.L., Wang, L., You, Z., Lou, Z.: MMSET regulates histone H4K20 methylation and 53BP1 accumulation at DNA damage sites. Nature **470**(7332), 124–128 (2011)
26. Pommier, Y., Sordet, O., Rao, V.A., Zhang, H., Kohn, K.W.: Targeting chk2 kinase: molecular interaction maps and therapeutic rationale. Curr. Pharm. Des. **11**(22), 2855–72 (2005)
27. Ray, O., Kakas, A.: ProLogICA: a practical system for abductive logic programming. In: Proceedings of the 11th International Workshop on Non-monotonic Reasoning, pp. 304–312 (2006)
28. Ray, O., Whelan, K., King, R.: Logic-based steady-state analysis and revision of metabolic networks with inhibition. In: CISIS'10, pp. 661–666 (2010)

29. Reiser, P.G., King, R.D., Kell, D.B., Muggleton, S., Bryant, C.H., Oliver, S.G.: Developing a logical model of yeast metabolism. Electron. Trans. Artif. Intell. **5**, 233–244 (2001)
30. Rougny, A., Froidevaux, C., Yamamoto, Y., Inoue, K.: Analyzing SBGN-AF Networks Using Normal Logic Programs. In: Logical Modeling of Biological Systems, pp. 44–55. Wiley, New York (2014)
31. van Iersel, M.P., Kelder, T., Pico, A.R., Hanspers, K., Coort, S., Conklin, B.R., Evelo, C.: Presenting and exploring biological pathways with pathvisio. BMC Bioinform., p. 399 (2008)

Efficient Unfolding of Fuzzy Connectives for Multi-adjoint Logic Programs

Pedro J. Morcillo and Ginés Moreno

Abstract During the last decade we have designed several tools for assisting the development of flexible software applications coded with a promising language in the fuzzy logic programming area. In the so-called *multi-adjoint logic programming* approach, a set of logic rules are assembled with a set of fuzzy connective definitions (whose truth functions are defined as functional rules) for manipulating truth degrees beyond the simpler case of *{true,false}*. Moreover, we have recently provided optimization techniques by reusing some variants of program transformation techniques based on unfolding which have been largely exploited in the pure functional -not fuzzy- setting for enhancing the behavior of such operators. In this paper we experimentally show the benefits of using the new *c-unfolding* transformation applied on fuzzy connectives and how to improve the efficiency of the proper unfolding process by reusing the very well-known concept of dependency graph. Moreover, we accompany our technique with cost analysis and discussions on practical aspects.

Keywords Fuzzy Logic Programming · Connectives · Unfolding

1 Introduction

Although *logic programming* [24] has been widely used as a formal method for problem solving and knowledge representation, *fuzzy logic programming* has emerged as a growing research area for incorporating techniques or constructs based on *fuzzy logic*

This work has been partially supported by the EU (FEDER), the State Research Agency (AEI) and the Spanish *Ministerio de Economía y Competitividad* under grant TIN2016-76843-C4-2-R (AEI/FEDER, UE).

P. J. Morcillo (✉) · G. Moreno
Department Computing System,
University of Castilla-La Mancha, 02071 Albacete, Spain
e-mail: pmorcillo@dsi.uclm.es

G. Moreno
e-mail: gines.moreno@uclm.es

© Springer International Publishing AG 2018
L. T. Kóczy and J. Medina (eds.), *Interactions Between Computational Intelligence and Mathematics*, Studies in Computational Intelligence 758,
https://doi.org/10.1007/978-3-319-74681-4_5

to explicitly deal with uncertainty and approximated reasoning. Most fuzzy logic languages developed during the last decades implement (extended versions of) the resolution principle introduced by Lee [21], such as Elf-Prolog [14], F-Prolog [23], generalized annotated logic programming [19], Fril [7], MALP [27], FASILL [15], the QLP scheme of [37] and the many-valued logic programming language of [38].

In this paper we focus on the so-called *multi-adjoint logic programming* approach MALP [25–27], a powerful and promising proposal in the area of fuzzy logic programming for which we have developed the $\mathcal{FLOPER}$ system (see [31, 36] and visit the Web site http://dectau.uclm.es/floper/) as well as several techniques for developing, optimizing and tuning fuzzy applications [10, 17, 33]. Intuitively speaking, logic programming is extended with a *multi-adjoint lattice L* of truth values (typically, a real number between 0 and 1), equipped with a collection of *adjoint pairs* $\langle \&_i, \leftarrow_i \rangle$ and connectives: implications, conjunctions, disjunctions, and other operators called aggregators, which are interpreted on this lattice.

In Sect. 2 we explain both the syntax and operational semantics of the MALP language where, in essence, to solve a MALP goal, i.e., a query to the system plus a substitution (initially the empty substitution, denoted by id), a generalization of the classical *modus ponens* inference rule called *admissible steps* are systematically applied on atoms in a similar way to classical resolution steps in pure logic programming, thus returning a state composed by a computed substitution together with an expression where all atoms have been exploited. Next, this expression is interpreted under a given lattice, hence returning a pair $\langle truth\ degree;\ substitution \rangle$ which is the fuzzy counterpart of the classical notion of computed answer used in pure logic programming.

In Sect. 3 we collect from [29] our technique for reducing the complexity of connectives (also alleviating the computational cost of derivations) by safely removing all the intermediate calls performed on the equations defining the behavior of such connectives. We show that this process can be easily described in terms of "unfolding", a well-known, widely used, semantics-preserving program transformation operation which in most declarative paradigms is usually based on the application of computation steps on the body of program rules (in [10, 16, 17] we describe our experiences regarding the unfolding of fuzzy logic programs). The novelty of our approach is that it is the first time that unfolding is not applied to program rules, but to connective definitions, maintaining the same final goal, i.e., generating more efficient code. The main three goals addressed in this paper are:

1. The advantages of the unfolding transformation adapted to connective definitions in our fuzzy setting are experimentally checked at the end of Sect. 3.
2. Moreover, we reuse in Sect. 4 some techniques based on dependency graphs for improving the proper transformation process as much as possible, also including computational cost analysis for our resulting algorithm.
3. In Sect. 5 we discuss some practical issues by connecting our techniques with several tools implemented in our research group.

Finally, before concluding in Sect. 7, a few hints on related work—which also help to motivate our approach—are presented in Sect. 6.

2   Multi-adjoint Logic Programming

This section summarizes the main features of multi-adjoint logic programming (see [25–27] for a complete formulation of this framework). We work with a first order language, $\mathcal{L}$, containing variables, constants, function symbols, predicate symbols, and several (arbitrary) connectives to increase language expressiveness: implication connectives ($\leftarrow_1, \leftarrow_2, \dots$); conjunctive operators (denoted by $\&_1, \&_2, \dots$), disjunctive operators ($|_1, |_2, \dots$), and hybrid operators (usually denoted by $@_1, @_2, \dots$), all of them are grouped under the name of "aggregators" or directly "connectives". Aggregation operators are useful to describe/specify user preferences. An aggregation operator, when interpreted as a truth function, may be an arithmetic mean, a weighted sum or in general any monotone application whose arguments are values of a complete bounded lattice L. For example, if an aggregator $@$ is interpreted as $[\![@]\!](x, y, z) = (3x + 2y + z)/6$, we are giving the highest preference to the first argument, then to the second, being the third argument the least significant. Although these connectives are binary operators, we usually generalize them as functions with an arbitrary number of arguments. So, we often write $@(x_1, \dots, x_n)$ instead of $@(x_1, \dots, @(x_{n-1}, x_n), \dots)$. By definition, the truth function for an n-ary aggregation operator $[\![@]\!] : L^n \to L$ is required to be monotonous and fulfills $[\![@]\!](\top, \dots, \top) = \top$, $[\![@]\!](\bot, \dots, \bot) = \bot$.

Additionally, our language $\mathcal{L}$ contains the values of a multi-adjoint lattice, $\langle L, \preceq, \leftarrow_1, \&_1, \dots, \leftarrow_n, \&_n \rangle$, equipped with a collection of adjoint pairs $\langle \leftarrow_i, \&_i \rangle$, where each $\&_i$ is a conjunctor which is intended to the evaluation of *modus ponens* [27]. In general, L may be the carrier of any complete bounded lattice but, for readability reasons, in the examples we shall select L as the set of real numbers in the interval $[0, 1]$. A *L-expression* is a well-formed expression composed by values and connectives of L, as well as variable symbols and *primitive operators* (i.e., arithmetic symbols such as $*, +, min$, etc.). In what follows, we assume that the truth function of any connective $@$ in L is given by its corresponding *connective definition*, that is, an equation or *rewriting rule* of the form $@(x_1, \dots, x_n) = E$, where E is a L-expression not containing variable symbols apart from $x_1, \dots, x_n$.

A *rule* is a formula $H \leftarrow_i \mathcal{B}$, where H is an atomic formula or atom (usually called the *head*) and $\mathcal{B}$ (which is called the *body*) is a formula built from atomic formulas $B_1, \dots, B_n$—$n \geq 0$—, truth values of L, conjunctions, disjunctions and aggregations. A *goal* is a body submitted as a query to the system. Roughly speaking, a multi-adjoint logic program is a set of pairs $\langle \mathcal{R}; \alpha \rangle$ (we often write $\mathcal{R}$ *with* α), where $\mathcal{R}$ is a rule and α is a *truth degree* (a value of L) expressing the confidence of a programmer in the truth of the rule $\mathcal{R}$. By abuse of language, we sometimes refer a tuple $\langle \mathcal{R}; \alpha \rangle$ as a "rule".

The procedural semantics of the multi-adjoint logic language $\mathcal{L}$ can be thought as an operational phase (based on admissible steps) followed by an interpretive one. In the following, $\mathcal{C}[A]$ denotes a formula where A is a sub-expression which occurs in the—possibly empty—context $\mathcal{C}[]$. Moreover, $\mathcal{C}[A/A']$ means the replacement of A by A' in context $\mathcal{C}[]$, whereas $\mathcal{V}ar(s)$ refers to the set of distinct variables occurring in the syntactic object s, and $\theta[\mathcal{V}ar(s)]$ denotes the substitution obtained from θ by restricting its domain to $\mathcal{V}ar(s)$.

Definition 1 *(Admissible Step)* Let $\mathcal{Q}$ be a goal and let σ be a substitution. The pair $\langle \mathcal{Q}; \sigma \rangle$ is a *state* and we denote by $\mathcal{E}$ the set of states. Given a program $\mathcal{P}$, an *admissible computation* is formalized as a state transition system, whose transition relation $\overset{AS}{\rightsquigarrow} \subseteq (\mathcal{E} \times \mathcal{E})$ is the smallest relation satisfying the following *admissible rules* (where we always consider that A is the selected atom in $\mathcal{Q}$ and $mgu(E)$ denotes the *most general unifier* of an equation set E):

(1) $\langle \mathcal{Q}[A]; \sigma \rangle \overset{AS}{\rightsquigarrow} \langle (\mathcal{Q}[A/v \&_i \mathcal{B}])\theta; \sigma\theta \rangle$
      if $\theta = mgu(\{A' = A\})$, $\langle A' \leftarrow_i \mathcal{B}; v \rangle$ in $\mathcal{P}$ and $\mathcal{B}$ is not empty.

(2) $\langle \mathcal{Q}[A]; \sigma \rangle \overset{AS}{\rightsquigarrow} \langle (\mathcal{Q}[A/v])\theta; \sigma\theta \rangle$
      if $\theta = mgu(\{A' = A\})$ and $\langle A' \leftarrow_i; v \rangle$ in $\mathcal{P}$.

As usual, rules are taken renamed apart. We shall use the symbols $\overset{AS1}{\rightsquigarrow}$ and $\overset{AS2}{\rightsquigarrow}$ to distinguish between computation steps performed by applying one of the specific admissible rules. Also, the application of a rule on a step will be annotated as a superscript of the $\overset{AS}{\rightsquigarrow}$ symbol.

Definition 2 Let $\mathcal{P}$ be a program and let $\mathcal{Q}$ be a goal. An *admissible derivation* is a sequence $\langle \mathcal{Q}; id \rangle \overset{AS}{\rightsquigarrow}* \langle \mathcal{Q}'; \theta \rangle$. When $\mathcal{Q}'$ is a formula not containing atoms (i.e., a *L*-expression), the pair $\langle \mathcal{Q}'; \sigma \rangle$, where $\sigma = \theta[\mathcal{V}ar(\mathcal{Q})]$, is called an *admissible computed answer* (a.c.a.) for that derivation.

Example 1 Let $\mathcal{P}$ be the following multi-adjoint logic program:

$$\mathcal{R}_1 : p(X) \leftarrow_{\texttt{godel}} \&_{\texttt{prod}}(|_{\texttt{luka}}(q(X), 0.6), r(X)) \text{ with } 0.9$$
$$\mathcal{R}_2 : q(a) \leftarrow \qquad\qquad\qquad\qquad\qquad\qquad\qquad \text{with } 0.8$$
$$\mathcal{R}_3 : r(X) \leftarrow \qquad\qquad\qquad\qquad\qquad\qquad\qquad \text{with } 0.7$$

where the labels `luka`, `godel` and `prod` mean respectively for *Łukasiewicz logic*, *Gödel logic* and *product logic*, that is,

$$|_{\texttt{luka}}(x_1, x_2) \quad = min(1, x_1 + x_2)$$
$$\&_{\texttt{prod}}(x_1, x_2) \quad = x_1 * x_2$$
$$\leftarrow_{\texttt{godel}}(x_1, x_2) = \; if \; (x_1 > x_2) \; then \; x_2 \; else \; 1$$
$$\&_{\texttt{godel}}(x_1, x_2) \quad = min(x_1, x_2)$$

Now, we can generate the following admissible derivation (we underline the selected atoms in each step):

$$\langle p(X);\ id \rangle \qquad\qquad\qquad\qquad\qquad\qquad\qquad\qquad \overset{AS1}{\leadsto}{}^{\mathcal{R}_1}$$

$$\langle \&_{\mathrm{godel}}(0.9, \&_{\mathrm{prod}}(|_{\mathrm{luka}}(q(X_1), 0.6), r(X_1))); \{X/X_1\} \rangle \qquad \overset{AS2}{\leadsto}{}^{\mathcal{R}_2}$$

$$\langle \&_{\mathrm{godel}}(0.9, \&_{\mathrm{prod}}(|_{\mathrm{luka}}(0.8, 0.6), \underline{r(a)})); \{X/a, X_1/a\} \rangle \qquad \overset{AS2}{\leadsto}{}^{\mathcal{R}_3}$$

$$\langle \&_{\mathrm{godel}}(0.9, \&_{\mathrm{prod}}(|_{\mathrm{luka}}(0.8, 0.6), \overline{0.7})); \{X/a, X_1/a, X_2/a\} \rangle$$

Here, the admissible computed answer (a.c.a.) is the pair: $\langle \&_{\mathrm{godel}}(0.9, \&_{\mathrm{prod}}(|_{\mathrm{luka}}(0.8, 0.6), 0.7)); \{X/a\} \rangle$.

If we exploit all atoms of a goal, by applying admissible steps as much as needed during the operational phase, then it becomes a formula with no atoms (a L-expression) which can be then directly interpreted w.r.t. lattice L. Although in [27] this last process is implicitly included in a definition similar to the previous one for describing the intended notion of fuzzy computed answer, here we prefer to model it as a new computational process (transition system) by applying the following definition we initially presented in [18] (as we will see in further sections, the cost measures proposed in this paper and previous ones [30, 32], are strongly related with the behaviour and detailed definition of interpretive step):

Definition 3 *(Interpretive Step)* Let $\mathcal{P}$ be a program, $\mathcal{Q}$ a goal and σ a substitution. We formalize the notion of *interpretive computation* as a state transition system, whose transition relation $\overset{IS}{\leadsto} \subseteq (\mathcal{E} \times \mathcal{E})$ is defined as the least one satisfying: $\langle Q[@(r_1, \ldots, r_n)]; \sigma \rangle \overset{IS}{\leadsto} \langle Q[@(r_1, .., r_n)/[\![@]\!](r_1, .., r_n)]; \sigma \rangle$, where $[\![@]\!]$ is the truth function of connective $@$ in the lattice $\langle L, \preceq \rangle$ associated to $\mathcal{P}$.

Definition 4 Let $\mathcal{P}$ be a program and $\langle Q; \sigma \rangle$ an a.c.a., that is, $\mathcal{Q}$ is a goal not containing atoms (i.e., a L-expression). An *interpretive derivation* is a sequence $\langle Q; \sigma \rangle \overset{IS}{\leadsto}* \langle Q'; \sigma \rangle$. When $Q' = r \in L$, being $\langle L, \preceq \rangle$ the lattice associated to $\mathcal{P}$, the state $\langle r; \sigma \rangle$ is called a *fuzzy computed answer* (f.c.a.) for that derivation.

Example 2 We complete the previous derivation of Example 1 by applying 3 interpretive steps in order to obtain the final f.c.a. $\langle 0.7; \{X/a\} \rangle$, thus generating the following interpretive derivation D_1:

$$\langle \&_{\mathrm{godel}}(0.9, \&_{\mathrm{prod}}(|_{\mathrm{luka}}(0.8, 0.6), 0.7)); \{X/a\} \rangle \quad \overset{IS}{\leadsto}$$

$$\langle \&_{\mathrm{godel}}(0.9, \&_{\mathrm{prod}}(1, 0.7)); \{X/a\} \rangle \qquad\qquad \overset{IS}{\leadsto}$$

$$\langle \&_{\mathrm{godel}}(0.9, 0, 7); \{X/a\} \rangle \qquad\qquad\qquad\qquad \overset{IS}{\leadsto}$$

$$\langle 0.7; \{X/a\} \rangle.$$

We have implemented the previous procedural principle into $\mathcal{FLOPER}$ as shown in Fig. 1, where each state (containing its associated goal and substitution) is colored in yellow and computational steps appear in blue circles (each admissible step is labeled with the used program rule, whereas label "is" reflects interpretive steps).

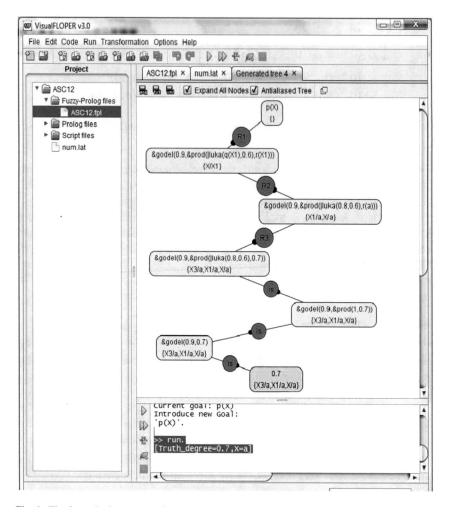

Fig. 1 The fuzzy logic programming environment $\mathcal{FLOPER}$

3 Unfolding Connective Definitions

As we said in the previous section, connective definitions are equations (or *rewriting rules*) of the form $@(x_1, \ldots, x_n) = E$, where E is a L-expression which might contain variable symbols in the set $\{x_1, \ldots, x_n\}$, as well as values, primitive operators and connectives of a multi-adjoint lattice L. The use of connectives inside the definition of other connectives is a powerful expressive resource useful not only for programmers interested in describing complex aggregators, but it also plays an important role in fuzzy transformation techniques such as the fold/unfold framework we have described in [10, 16–18]. Consider for instance, the following connective definition: $@_{\text{complex}}(x_1, x_2) = \&_{\text{prod}}(|_{\text{luka}}(x_1, 0.6), x_2)$. This hybrid aggregator was used in [30, 32] (with slight modifications) for pointing out some observed discrepancies when measuring the interpretive cost associated to the execution of MALP programs.

Example 3 A simplified version of rule $\mathcal{R}_1$ in Example 1, whose body only contains an aggregator symbol is $\mathcal{R}_4 : p(X) \leftarrow_{\text{godel}} @_{\text{complex}}(q(X), r(X))$ with 0.9. Note that $\mathcal{R}_4$ has exactly the same meaning (interpretation) that $\mathcal{R}_1$, although different syntax. In fact, both rules have the same sequence of atoms in their head and bodies. The differences are the set of connectives which explicitly appear in their bodies since in $\mathcal{R}_4$ we have moved $\&_{\text{P}}$ and $|_{\text{luka}}$ (as well as value 0.6) from the body of the rule (see $\mathcal{R}_1$) to the connective definition of $@_{\text{complex}}$.

On the other hand, a classical, simple way for estimating the computational cost required to built a derivation, consists in counting the number of computational steps performed on it. So, given a derivation D, we define:

– *operational cost* $\mathcal{O}_c(D)$, as the number of admissible steps performed in D, and
– *interpretive cost* $\mathcal{I}_c(D)$, as the number of interpretive steps done in D.

Note that the operational and interpretive costs of derivation D_1 performed in the previous section are $\mathcal{O}_c(D_1) = 3$ and $\mathcal{I}_c(D_1) = 3$, respectively. Intuitively, $\mathcal{O}_c$ informs us about the number of atoms exploited along a derivation. Similarly, $\mathcal{I}_c$ seems to estimate the number of connectives evaluated in a derivation. However, this last statement is not completely true: $\mathcal{I}_c$ only takes into account those connectives appearing in the bodies of program rules which are replicated on states of the derivation, but no those connectives recursively *nested* in the definition of other connectives, as we are going to see. Let us use rule $\mathcal{R}_4$ instead of $\mathcal{R}_1$ for generating the following derivation D_1^* which returns the same f.c.a than D_1:

$$\langle p(X); \ id \rangle \qquad\qquad\qquad\qquad\qquad\qquad\qquad \overset{\mathcal{R}_4}{\underset{AS1}{\rightsquigarrow}}$$

$$\langle \&_{\text{godel}}(0.9, @_{\text{complex}}(q(X_1), r(X_1)); \{X/X_1\} \rangle \qquad \overset{\mathcal{R}_2}{\underset{AS2}{\rightsquigarrow}}$$

$$\langle \&_{\text{godel}}(0.9, @_{\text{complex}}(0.8, r(a))); \{X/a, X_1/a\} \rangle \qquad \overset{\mathcal{R}_3}{\underset{AS2}{\rightsquigarrow}}$$

$$\langle \&_{\text{godel}}(0.9, @_{\text{complex}}(0.8, 0.7)); \{X/a, X_1/a, X_2/a\} \rangle \qquad \overset{IS}{\rightsquigarrow}$$

$$\langle \&_{\text{godel}}(0.9, 0.7); \{X/a, X_1/a, X_2/a\} \rangle \qquad\qquad\qquad \overset{IS}{\rightsquigarrow}$$

$$\langle 0.7; \{X/a, X_1/a, X_2/a\} \rangle$$

Note that, since we have exploited the same atoms with the same rules (except for the first steps performed with the equivalent rules $\mathcal{R}_1$ and $\mathcal{R}_4$, respectively) in both derivations, then $\mathcal{O}_c(D_1) = \mathcal{O}_c(D_1^*) = 3$. However, although connectives $\&_{\text{prod}}$ and $|_{\text{luka}}$ have been evaluated in both derivations, in D_1^* such evaluations have not been explicitly counted as interpretive steps, and consequently they have not been added to increase the interpretive cost measure $\mathcal{I}_c$. This unrealistic situation is reflected by the abnormal result: $\mathcal{I}_c(D_1) = 3 > 2 = \mathcal{I}_c(D_1^*)$. In [30, 32] we have described two different techniques (based respectively on a redefinition of the notion of interpretive step and in the introduction of the concept of "weight of a connective") evidencing that the interpretive cost of derivation D_1^* is not only lower, but even greater than derivation D_1. The main reason is that complex connective definitions involving calls to other aggregators consume more computational resources than other connectives which only evaluate primitive operators.

The previous example motivates the following definition, which in essence describes a technique based on classical unfolding transformations for simplifying, when possible, connective definitions by "unnesting" unnecessary calls to other connectives.

Definition 5 *(C-Unfolding)* Let $\langle L, \preceq \rangle$ be a multi-adjoint lattice containing the connective definitions $@(x_1, \ldots, x_n) = E$ and $@'(x_1', \ldots, x_m') = E'$, such that a call to $@'$ of the form $@'(t_1, \ldots, t_m)$ appears in E. Then, the unfolding of connective $@$ w.r.t. connective $@'$ or directly, the c-unfolding of $@$, is the new equation: $@(x_1, \ldots, x_n) = E[@'(t_1, \ldots, t_m)/E'']$, where E'' is obtained from the L-expression E' by replacing each variable (formal parameter) x_i' by its corresponding value (actual parameter) t_i, $1 \leq i \leq m$, that is $E'' = E'[x_1'/t_1, \ldots, x_m'/t_m]$.

We assume here that the rules (equations) describing connective definitions are taken renamed apart (at least one of them) before applying an unfolding step, as it is also usual with program rules in many declarative transformation tasks.

Example 4 Focusing now in the connective definition:

$$@_{\text{complex}}(x_1, x_2) = \&_{\text{prod}}(|_{\text{luka}}(x_1, 0.6), x_2)$$

... and remembering that $|_{\text{luka}}(x_1', x_2') = min(1, x_1' + x_2')$, then, we can unfold connective $@_{\text{complex}}$ w.r.t. connective $|_{\text{luka}}$ as follows:

- Firstly, we generate the "matcher" between the call $|_{\text{luka}}(x_1, 0.6)$ appearing in the "rhs" (right hand side) of the first equation rule and the "lhs" (left hand side) of the second rule $|_{\text{luka}}(x_1', x_2')$, thus producing links x_1'/x_1 and $x_2'/0.6$.
- Next, we apply both bindings to the rhs of the second rule, obtaining the L-expression $min(1, x_1 + 0.6)$.
- Then, this L-expression is used to replace the original call to $|_{\text{luka}}$ in the rhs of the first rule, producing $\&_{\text{prod}}(min(1, x_1 + 0.6), x_2)$.
- Finally, this last L-expression conforms the rhs of the new connective definition for $@_{\text{complex}}$, that is: $@_{\text{complex}}(x_1, x_2) = \&_{\text{prod}}(min(1, x_1 + 0.6), x_2)$.

Following the same method, but performing now the c-unfolding of $@_{\text{complex}}$ w.r.t. $\&_{\text{prod}}$ whose connective definition is $\&_{\text{prod}}(x_1, x_2) = x_1 * x_2$, we obtain the final rule defining $@_{\text{complex}}$ with the following shape $@_{\text{complex}}(x_1, x_2) = min(1, x_1 + 0.6) * x_2$. Note that the new connective definition is just a simple arithmetic expressions involving primitive operators but no calls to other connectives, as wanted. From now on, this improved definition will be referred as $@_{\text{unfolded}}$.

To finish this section, in Table 1 we show the benefits of using c-unfolding by means of a experimental evaluation performed on a desktop computer equipped with an

Table 1 Evaluating original and unfolded fuzzy connectives: runtimes and speed-up

MN	100	1000	10,000
100	0.04 / 0.03 = 1.33	0.31 / 0.3 = 1.03	2.91 / 2.89 = 1.01
1000	0.11 / 0.03 = 3.67	0.38 / 0.3 = 1.27	3.03 / 2.93 = 1.03
10000	0.76 / 0.04 = 19	1.02 / 0.3 = 3.4	3.62 / 2.92 = 1.24

i3-2310M CPU @ 2.10 GHz and 4,00 GB RAM. We consider an initial complex connective whose definition contains (in a direct or indirect way) N calls to other connectives and requires the evaluation of M primitive operators like min, $*$, $+$, and so on. More exactly, assuming that a connective @ directly evaluates m primitive operators and performs n direct calls to connectives $@_1, \ldots, @_n$, then, we can compute values M and N by means of the auxiliary functions $opers$ and $calls$ as follows: M $= opers(@) = m + opers(@_1) + \cdots + opers(@_n)$ and N $= calls(@) = n + calls(@_1) + \cdots + calls(@_n)$. For instance, if we try to compute the values M and N for connective $@_{\text{complex}}(x_1, x_2) = \&_{\text{prod}}(|_{\text{luka}}(x_1, 0.6), x_2)$ in Example 4, we have that $m = 0, n = 2$ and since $|_{\text{luka}}(x_1', x_2') = min(1, x_1' + x_2')$ and $\&_{\text{prod}}(x_1, x_2) = x_1 * x_2$, then $opers(|_{\text{luka}}) = 2$ and $opers(\&_{\text{prod}}) = 1$, while $calls(|_{\text{luka}}) = calls(\&_{\text{prod}}) = 0$, which implies that M $= opers(@_{\text{complex}}) = m + opers(|_{\text{luka}}) + opers(\&_{\text{prod}}) = 0 + 2 + 1 = 3$ and N $= calls(@_{\text{complex}}) = n + calls(|_{\text{luka}}) + calls(\&_{\text{prod}}) = 2 + 0 + 0 = 2$. Remember that after applying c-unfolding, the improved connective definition does not perform calls to any other connective, but evaluates the same number of primitive operators M (in fact, observe that in the improved definition $@_{\text{complex}}(x_1, x_2) = (min(1, x_1 + 0.6)) * x_2$ of Example 4 we have that M = 3 and N = 0, as wanted).

Each row in Table 1 refers to a different value for N while each column indicates an alternative assignment to M. Both parameters vary according to values 100, 1000 and 10,000. Each case has been executed 1000 times and the content of each cell has the form "runtime-original-connective / runtime-unfolded-connective = speed-up", where execution times are expressed in milliseconds. Note that the speed up in the cells of the first row are not significant due to the fact that the number of connective calls N is never greater than the number of primitive operators M. On the contrary, in the last row, since N is always greater or equal than M, we obtain good ranges of speed up. In particular, this measure is 19 in the leftmost cell due to the fact that c-unfolding removes all the connective calls which caused the low efficiency of the initial connective definition.

4 C-Unfolding and Dependency Graphs

The use of "graphs" (and many different extensions/variants of this formal concept) as an auxiliary data structure helping to analyze the behaviour of systems/programs at several levels, also taking profit in practice of its deep mathematical background. For instance, and simply focusing on termination topics in declarative programming (which has somehow influenced our recent research interest), the notions of *dependency graphs* and *size-change graphs* have been well reported in [5, 22]. In this section, we will use the first concept for efficiently guiding the c-unfolding process.

Our experiences in fuzzy fold/unfold transformations [10, 16, 17], reveal us that drastic situations associated to degenerated transformation sequences might eventually produce highly nested definitions of connectives. For instance, assume the following sequence of (extremely inefficient) connective definitions:

$$@_{100}(x_1, x_2) = @_{99}(x_1, x_2)$$
$$@_{99}(x_1, x_2) = @_{98}(x_1, x_2)$$
$$@_{98}(x_1, x_2) = @_{97}(x_1, x_2)$$
$$........ \qquad\qquad ....$$
$$@_1(x_1, x_2) = @_0(x_1, x_2)$$
$$@_0(x_1, x_2) = x_1 * x_2$$

When trying to solve two expressions of the form $@_{100}(0.9, 0.8)$ and $@_0(0.9, 0.8)$, we obtain the same result 0.72, but the effort needed to solve the first expression is very high (due to the 100 avoidable calls to auxiliary connectives) compared with the second expression (which simply evaluates the arithmetic operator $*$).

Fortunately, by systematically performing c-unfolding on the previous connectives, this problem is successfully solved in a simple way: after applying a c-unfolding step on aggregator $@_{100}$ we obtain $@_{100}(x_1, x_2) = @_{98}(x_1, x_2)$, which admits a new c-unfolding process to become $@_{100}(x_1, x_2) = @_{97}(x_1, x_2)$, and following this trail, after applying the final one-hundredth c-unfolding step, we reach the desired connective definition $@_{100}(x_1, x_2) = x_1 * x_2$. Of course, the transformation process does not finish here, because we also need to rearrange the shape of all the remaining connective definitions. So, for each aggregator $@_i$, $0 \leq i \leq 100$, we need exactly i c-unfolding steps to achieve the appropriate connective definition.

However, there exist a second, much more intelligent alternative to highly reduce the number of transformation steps needed to obtain the same final set of improved connective definitions. In our example, the idea is to proceed just in the inverse order than previously. So, since $@_0$ does not admit unfolding, we proceed with $@_1$, whose connective definition becomes $@_1(x_1, x_2) = x_1 * x_2$ after just a single c-unfolding step. Now, we take profit of this improved definition when unfolding $@_2$, since in just a unique (not two) c-unfolding step we obtain the optimal definition $@_2(x_1, x_2) = x_1 * x_2$. Note that the benefits of this last process, are also inherited when transforming $@_3$, $@_4$ and so on. So, the advantages obtained after applying each c-unfolding on a different connective, are "propagated" to the remaining connectives being improved, which implies that we simply need one hundred transformation steps to optimize the definitions of the whole set of connectives.

In order to identify in a systematic way the best ordering for performing c-unfolding operations on connectives, we firstly construct the *dependency graph* of a multi-adjoint lattice L associated to a given program $\mathcal{P}$, i.e., a directed graph that contains the connective symbols as nodes and an edge from connective @ to aggre-gator @' for each connective definition in L of the form $@(x_1, \ldots, x_n) = E$, where the L-expression E contains a call to @'. Given an edge from node @ to node @', we denote it as an *out-edge* of @ and as an *in-edge* of @'. For instance, the dependency graphs associated to all the connectives seen so far are:

$$\boxed{\vee_L} \longleftarrow \boxed{@^*} \longrightarrow \boxed{\&_P}$$

$$\boxed{@_{100}} \longrightarrow \boxed{@_{99}} \longrightarrow \ldots \longrightarrow \boxed{@_1} \longrightarrow \boxed{@_0}$$

As we are going to see, the use of dependency graphs will largely help us to decide when to unfold each connective in order to minimize the number of transformation steps. Anyway, before doing this, it is important to note that the construction of such graphs constitute a fast way to detect possibly abnormal connective definitions, that is, those ones involved on cycles in the graph (because their further evaluation might fall in infinite loop). Fortunately, the presence of cycles is not usual in the dependency graphs associated to connective definitions.

   As formalized in the algorithm of Fig. 2, when selecting a connective to apply c-unfolding, we give priority to those ones without out-edges (and obviously not belonging to cycles), as occurs in our examples with nodes labeled with $\vee_L$, $\&_P$ and $@_0$, which in our particular case do not need c-unfolding because their definitions do not perform calls to other aggregators. Once a concrete connective has been selected and then unfolded as much as possible (and hence, its definition has been completely improved by removing all its auxiliary calls), then the proper node as well as all its in-edges (remember that it has not associated out-edges) are removed from the graph. The process is iterated as much as needed until the dependency graph becomes empty. For instance in our example, once removed nodes $\vee_L$, $\&_P$ and $@_0$, the new candidates are nodes $@^*$ and $@_1$. The first one is unfolded w.r.t. $\vee_L$ and $\&_P$ and then removed, whereas the second one is dropped out after being unfolded w.r.t. $@_0$. Then the process continues with $@_2$, next $@_3$ and so on, being $@_{100}$ the last

```
INPUT: Multi-adjoint lattice L with nested connective definitions.

BODY:
  1. Build a "dependency graph" G from L, where:
     (a) each node in G represents a connective defined in L.
     (b) each directed edge in G "means" a call from @ to @'.
  2. While G is not empty:
     (a) select a node/connective @ in G without out-edges,
     (b) apply C-unfolding as much as possible on node/connective @,
     (c) remove @ as well as all its (in/out) edges from G.

OUTPUT: Multi-adjoint lattice L without nested connective definitions.
```

Fig. 2 Algorithm for efficiently unfolding connective definitions

connective whose definition is optimized by applying just a single c-unfolding step, thus accomplishing with the desired ordering which produces the benefits reported along this section.

Focusing on cost analysis, we simply wish to indicate that, for a given dependency graph with n nodes and m edges, it is obvious to see that the number of c-unfolding steps performed by our algorithm is exactly m, thanks to the high priority given to nodes without out-edges for being c-unfolded. On the contrary, if we firstly c-unfold nodes without in-edges, then the complexity of the algorithm could be greater than linear (even quadratic), as illustrated by our example at the beginning of this section.

5   Implementation and Practical Issues

In this section we establish the feedback and synergies among several independent tools developed in our research group during the last decade. Whereas in [9] we present a recent graphical tool for assisting the design of lattices of truth degrees (see Fig. 3), the functional-logic transformation system $\mathcal{SYNTH}$ enables the unfolding-based optimization of their connective definitions in order to improve the computational behaviour of those fuzzy logic programs developed with the $\mathcal{FLOPER}$ environment.

The transformation system $\mathcal{SYNTH}$ [4, 35] was initially conceived for optimizing functional logic programs (CURRY [11]), and then also used to manipulate pure functional programs (HASKELL [13]). The tool implements five basic transformation rules (unfolding, folding, abstraction, definition introduction and definition elimination) and two automatic transformation strategies (composition and tupling), in order to generate efficient sets of rewriting rules coded with HASKELL/CURRY syntax.[1] For the purposes of the present work, we simply need to consider the unfolding operation of the $\mathcal{SYNTH}$ system for being applied on rewriting rules modeling fuzzy connectives, as occurs with the following ones of our running example:

```
or_luka A B = min 1 (add A B)
and_prod A B = prod A B
agr_complex A B = and_prod (or_luka A (0,6)) B
```

Here, the connectives $|_{luka}$, $\&_{prod}$ and $@_{complex}$ are respectively denoted by or_luka, and_prod and agr_complex; we use the names min, add and prod for referring to the primitive arithmetic operators min, $+$ and $*$, respectively; and finally A and B are variable symbols. Once the previous set of connective definitions is loaded into the $\mathcal{SYNTH}$ system, it conforms the initial program in the transformation sequence denoted by "asc 0". Next, we must select the third rule and click on the unfolding button, thus obtaining the next program "asc 1"

---

[1] In essence, both languages share the same syntax, but they have a different computational behaviour since CURRY extends with extra logic features the pure functional dimension of HASKELL.

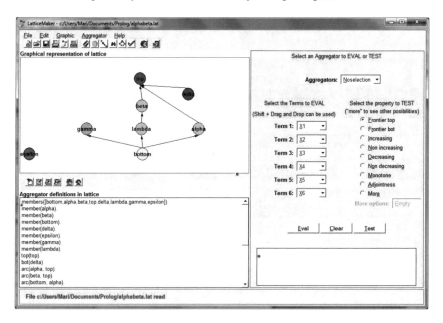

Fig. 3 Graphical design of lattices with the `Lattice-Maker` tool

where the selected rule is replaced by the new fourth one `agr_complex A B = and_prod (min 1 (add A (0,6)))` B. Finally, this last rule admits a final unfolding step to reach our intended final definition for `agr_complex` (where no calls to other connectives appear), represented by the fifth rule `agr_complex A B = prod (min 1 (add A (0,6)))` B in program "`asc 2`". Figure 4 shows the original and final program of the transformation sequence, where the initial and improved definitions of `agr_complex` appear darked in blue.

On the other hand, our "*Fuzzy LOgic Programming Environment for Research*" $\mathcal{FLOPER}$, is able to trace the execution of goals with respect to a given MALP program and its associated lattice, by drawing derivation trees as the ones shown in Figs. 1 and 5. When we choose option "`ismode = small`" then the system is able to detail the set of connective calls performed along a derivation (`sis1` steps) as well as the set of primitive operators (`sis2` steps) evaluated in the same derivation. Before explaining the three derivations collected on the tree drawn in Fig. 5, remember that our MALP program looks like:

$$\mathcal{R}_1 : p(X) \leftarrow_{\text{godel}} \&_{\text{prod}}(|_{\text{luka}}(q(X), 0.6), r(X)) \text{ with } 0.9$$
$$\mathcal{R}_2 : q(a) \leftarrow \qquad\qquad\qquad\qquad\qquad\qquad \text{with } 0.8$$
$$\mathcal{R}_3 : r(X) \leftarrow \qquad\qquad\qquad\qquad\qquad\qquad \text{with } 0.7$$
$$\mathcal{R}_4 : p(X) \leftarrow_{\text{godel}} @_{\text{complex}}(q(X), r(X)) \qquad \text{with } 0.9$$
$$\mathcal{R}_5 : p(X) \leftarrow_{\text{godel}} @_{\text{unfolded}}(q(X), r(X)) \qquad \text{with } 0.9$$

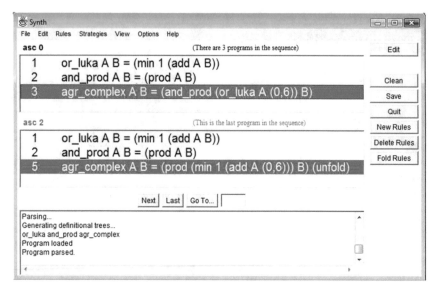

Fig. 4 The transformation system $\mathcal{SYNTH}$

where @$_{\text{unfolded}}$ represents the improved definition (where no calls to other connectives are performed) of @$_{\text{complex}}$.

In the figure, each derivation starts with an admissible step exploiting a different MALP rule defining predicate p (i.e., $\mathcal{R}_1$, $\mathcal{R}_4$ and $\mathcal{R}_5$, respectively) and continues with rules $\mathcal{R}_2$ and $\mathcal{R}_3$ in all cases until finishing the admissible phase. From here, the interpretive phase differs in each derivation since, even when all them evaluate exactly the same set of primitive operators (so, the number of sis2 steps do coincide), they perform a different number of direct/indirect calls to connectives (represented by sis1 steps). So, the left-most branch in Fig. 5 is shorter than the branch in the middle of the tree (as we have largely explained in the previous section), but note that the right-most branch, which makes use of the connective whose definition has been improved by means of our c-unfolding technique, is just the shortest one in the whole figure, as wanted.

We have just seen by means of a very simple example that the "quality" of the connective definitions accompanying a MALP program directly reverts on the efficiency of the applications coded with this fuzzy language. Thus, the unfolding technique applicable on fuzzy connectives described in this paper, is intended to play an important role in the performance of large scale programs developed with our $\mathcal{FLOPER}$ environment, as it is the case of the real-world application we have recently developed with $\mathcal{FLOPER}$ in the field of the semantic web [2, 3] (see Fig. 6).

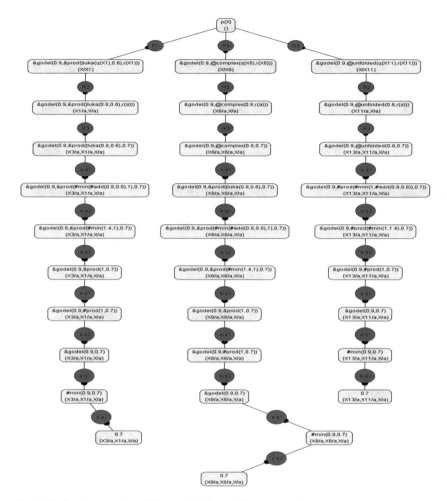

Fig. 5 Derivation tree using different definitions of fuzzy connectives

6 Related Work

The ideas managed in the previous sections have been mainly inspired by our previous studies and experiences in the following two topics: fold/unfold transformations in declarative programming and cost measures in multi-adjoint logic programming ([28, 30, 32]).

Focusing on primitive functional programs, the pioneer work [8] initiates a fertile tradition in program optimization techniques based on fold/unfold transformations, which has highly attracted a wide audience in the declarative programming research community during the last three decades (see the first introduction into logic programming in [39], and then our adaptations to functional logic programming

Fig. 6 An on-line work session with the `fuzzyXPath` application

in [4, 35] and fuzzy logic programming in [10, 16–18]). This approach is based on the construction, by means of a *strategy* (heuristic), of a sequence of equivalent programs—called *transformation sequence* and usually denoted by $\mathcal{P}_0, \ldots, \mathcal{P}_n$ such that $\mathcal{P}_n$ is more efficient than $\mathcal{P}_0$—where each program $\mathcal{P}_i$ is obtained from the preceding ones $\mathcal{P}_0, \ldots, \mathcal{P}_{i-1}$ by using an *elementary* transformation rule. The essential rules are *folding* and *unfolding*, i.e., contraction and expansion of sub-expressions of a program using the definitions of this program (or a preceding one).

Example 5 Consider for instance the following set of rewriting rules describing classical operations for concatenating two and three lists via the *app* and *double* function symbols, respectively (here we use the constructor symbols "[ ]" and ":" to model empty and non-empty lists, respectively):

$$\mathcal{R}_1: \quad app([\,], L) \quad = L$$
$$\mathcal{R}_2: \quad app(H:T, L) \quad = H: app(T, L)$$
$$\mathcal{R}_3: \quad double(A, B, C) = \underline{app(app(A, B), C)}$$

This is a classical example of program optimization by fold/unfold, where it is easy to see that the underlined expression in rule $\mathcal{R}_3$ is rather inefficient because its evaluation requires the double traversal of the first list (represented by variable A). Here, we can apply two unfolding steps on rule $\mathcal{R}_3$ until obtaining the new pair of rules:

$$\mathcal{R}_4: double([\,], B, C) \quad = app(B, C)$$
$$\mathcal{R}_6: double(H:A, B, C) = H: \underline{app(app(A, B), C)}$$

Now, observe that the underlined expression in $\mathcal{R}_6$ coincides with the body of the right hand side of rule $\mathcal{R}_3$, and thus, such expression admits a "contraction" in terms of "*double*". This is just what the final folding step produces, thus generating the new rule $\mathcal{R}_7 : double(H : A, B, C) = H : double(A, B, C)$. The reader can easily check that this recursive definition of "*double*" is much more efficient than the one proposed in the original program, since in order to concatenate three lists, only one traversal of the first two lists is required now. So, the transformation process successfully finishes with the following final program:

$$
\begin{array}{lll}
\mathcal{R}_1 : & app([\,], L) & = L \\
\mathcal{R}_2 : & app(H : T, L) & = H : app(T, L) \\
\mathcal{R}_4 : & double([\,], B, C) & = app(B, C) \\
\mathcal{R}_7 : & double(H : A, B, C) = H : double(A, B, C)
\end{array}
$$

On the other hand, in [10] we have developed a much more involved fuzzy fold/unfold transformation system especially tailored for the MALP framework, whose advantages were illustrated with the following program:

$$
\begin{array}{llll}
\mathcal{R}_1 : p(X, Y) & \leftarrow_{\texttt{luka}} q(X, Z) \&_{\texttt{godel}} q(Z, Y) & \text{with} & 0.93 \\
\mathcal{R}_2 : q(a, b) & \leftarrow & \text{with} & 0.9 \\
\mathcal{R}_3 : q(b, c) & \leftarrow & \text{with} & 0.85 \\
\mathcal{R}_4 : q(f(X), g(Y)) & \leftarrow_{\texttt{prod}} q(X, Y) & \text{with} & 0.8 \\
\mathcal{R}_5 : q(g(X), h(Y)) & \leftarrow_{\texttt{prod}} q(X, Y) & \text{with} & 0.95
\end{array}
$$

After applying the transformation process, we obtain the improved version:

$$
\begin{array}{llll}
\mathcal{R}_{11} : new(a, b, c) & \leftarrow_{\texttt{prod}} \langle 0.9, 0.85 \rangle & \text{with} & 1 \\
\mathcal{R}_{16} : new(f(X), g(Z), h(Y)) & \leftarrow_{\texttt{prod}} @_1(new(X, Z, Y)) & \text{with} & 1 \\
\mathcal{R}_{18} : p(X, Y) & \leftarrow_{\texttt{luka}} @_2(new(X, Z, Y)) & \text{with} & 0.93
\end{array}
$$

where the new aggregators used in the body of rules $\mathcal{R}_{16}$ and $\mathcal{R}_{18}$ are defined as $@_1(\langle x_1, x_2 \rangle) = \langle \&_{\texttt{prod}}(0.8, x_1), \&_{\texttt{prod}}(0.95, x_2) \rangle$ and $@_2(\langle x_1, x_2 \rangle) = \&_{\texttt{godel}}(x_1, x_2)$. In this fuzzy setting, apart from the folding/unfolding operations, it is mandatory to introduce two new transformations called "tupled definition introduction" and "aggregation", being this last operation strongly connected with the notions introduced in this paper, as we are going to see.

As in our previous example, Burstall and Darlington considered in [8] a set of equations defining functions (also using variables and constructor symbols to build data structures) as functional programs, with a shape very close to the connective definitions used in this paper, by simply associating their notions of function symbols and constructor symbols (or more properly, "defined function symbols" and "constructor function symbols", respectively), to our concepts of connective symbols and primitive operators, respectively. In fact, our equations describing connectives can be seen as a particular case of their equations defining program rules, since in the left hand sides of connective definitions we only allow variables as arguments (they can also use terms built with constructors).

For this reason, our notion of c-unfolding does not only accomplish with the original unfolding definition of [8] (and further, more complex extensions coping with modern functional and functional-logic programs expressed as *Term Rewriting Systems, TRS's* [6, 12, 20]), but it is even easier, since we simply need a restricted kind of equations to model connective definitions. Hence, apart from our initial, novel use of unfolding described in this paper, we think that in the near future it would be possible to extrapolate to our fuzzy setting many results on transformation techniques (including folding) with a *functional taste*, but focusing now on the optimization of connective definitions.

At this point, we think that it is mandatory to explain why in our initial fuzzy fold/unfold transformation system described in [10, 18, 34] there exists the risk of generating artificial and inefficient connective definitions when optimizing multi-adjoint logic programs. The problem emerges in the fuzzy unfolding transformation since this operation introduces a great amount of connectives and truth degrees on unfolded program rules which obstruct the application of further folding steps (as we have done in Example 5 when generating efficient recursive definitions).

Example 6 Such effects are easy to see even in the propositional case. So, in order to unfold a fuzzy program rule like:

$$p \leftarrow_{\text{luka}} q \ with \ 0.9$$

we must apply an admissible step (according to the procedural semantics described in Definition 1) on its body, by using for instance a rule like:

$$q \leftarrow_{\text{godel}} (r \ \&_{\text{prod}} s) \ with \ 0.8$$

Hence, we obtain the new unfolded rule:

$$p \leftarrow_{\text{luka}} (\underline{0.8 \ \&_{\text{godel}} (r \ \&_{\text{prod}} s)}) \ with \ 0.9$$

where the underlined elements confirm our previous comment, that is, the introduction of extra truth degrees and connectives on the body of unfolded rules.

It is important to note that the existence of such "noisy" elements only emerge in the fuzzy setting (in contrast with other "crisp" languages) and, for manipulating them, we require the application of auxiliary techniques which, as we are going to see, will produce artificial connective definitions: the main motivation of this paper is just the optimization of such connective definitions by following standard declarative techniques classically proposed for the optimization of program rules.

So, if we revise the so-called "aggregation transformation rule" described in [10], we observe that its main goal is to simplify the shape of program rules, by moving several connective calls from their bodies to the definition of new connective symbols, in order to give chances for further folding steps to proceed.

Example 7 Having a look to previous examples, the effects produced by the "aggregation transformation rule" are:

- The program rule initially introduced in Example 1

$$\mathcal{R}_1 : \ p(X) \leftarrow_{\texttt{godel}} \&_{\texttt{prod}}(|_{\texttt{luka}}(q(X), 0.6), r(X)) \text{ with } 0.9$$

becomes the transformed rule of Example 3

$$\mathcal{R}_1^* : \ p(X) \leftarrow_{\texttt{godel}} @^*(q(X), r(X)) \text{ with } 0.9$$

where remember that $@^*(x_1, x_2) = \&_{\texttt{prod}}(|_{\texttt{luka}}(x_1, 0.6), x_2)$.
- Regarding now Example 6, we observe that the program rule obtained after an unfolding step:

$$p \leftarrow_{\texttt{luka}} (0.8 \&_{\texttt{godel}} (r \&_{\texttt{prod}} s)) \ with \ 0.9$$

could be replaced by application of an "aggregation step" by the new rule:

$$p \leftarrow_{\texttt{luka}} @(r, s) \ with \ 0.9$$

where $@(x_1, x_2) = (0.8 \&_{\texttt{godel}} (x_1 \&_{\texttt{prod}} x_2))$.

Some important features of this transformation operation are:

- No folding step is allowed if a previous aggregation step has not been previously performed.
- Each aggregation step introduces a new connective definition which necessarily invokes other connectives previously known, thus producing nested definitions of aggregators.
- When optimizing programs in practice, several folding/unfolding steps are often required even on a same given program rule which, in our fuzzy setting, implies several applications of the aggregation rule in an iterative way (which reinforces the presence of nested connective definitions when fuzzy programs are manipulated with these techniques).

Anyway, it must be taken into account that as in many other declarative paradigms, the time spent once during the correct application of fold/unfold techniques for optimizing program rules, is largely compensated forever by the fast execution of the resulting refined programs. In the fuzzy setting, we have seen that this transformation process focusing only on program rules, also requires a second stage for rearranging the shape of such "artificial" connective definitions probably produced during the first transformation phase. The techniques reported along this paper achieve this last goal (just once, and in a finite time which is reduced as much as possible by the use of call graphs) on connective definitions without disturbing the benefits initially reached by fold/unfold on fuzzy program rules.

7 Conclusions and Future Work

In this paper we were concerned with the optimization of fuzzy logic connectives whose artificial, inefficient definitions could have been automatically produced by previous transformation processes applied on fuzzy MALP programs. Our technique, inspired by rewriting-based unfolding, takes profit from clear precedents in pure functional programming. In this paper we have focused on the optimization of the proper unfolding process (initially presented in [29]) by making use of dependency graphs in order to decide the ordering in which several connective calls must be unfolded inside a concrete connective definition. For the near future, we plan to implement our technique inside the fuzzy logic programming environment $\mathcal{FLOPER}$ (visit http://dectau.uclm.es/floper/) we have designed for developing applications coded with the MALP language [1, 3, 40].

References

1. Almendros-Jiménez, J.M., Bofill, M., Luna Tedesqui, A., Moreno, G., Vázquez, C., Villaret, M.: Fuzzy xpath for the automatic search of fuzzy formulae models. In: Beierle, C., Dekhtyar A. (eds.) Scalable Uncertainty Management—9th International Conference, SUM 2015, Québec City, QC, Canada, September 16–18, 2015. Proceedings, volume 9310 of Lecture Notes in Computer Science, 385–398. Springer, 2015
2. Almendros-Jiménez, J.M., Luna, A., Moreno, G.: Fuzzy logic programming for implementing a flexible xpath-based query language. Electron. Notes Theor. Comput. Sci. **282**, 3–18 (2012)
3. Almendros-Jiménez, J.M., Luna, A., Moreno, G.: Fuzzy xpath through fuzzy logic programming. New Generation Computing **33**(2), 173–209 (2015)
4. Alpuente, M., Falaschi, M., Moreno, G., Vidal, G.: Rules + strategies for transforming lazy functional logic programs. Theoretical Comput. Sci. Elsevier **311**(1–3), 479–525 (2004)
5. Arts, T., Giesl, J.: Termination of term rewriting using dependency pairs. Theor. Comput. Sci. **236**(1–2), 133–178 (2000)
6. Baader, F., Nipkow, T.: Term Rewriting and All That. Cambridge University Press, Cambridge (1998)
7. Baldwin, J.F., Martin, T.P., Pilsworth, B.W.: Fril- Fuzzy and Evidential Reasoning in Artificial Intelligence. Wiley, New Yark (1995)
8. Burstall, R.M., Darlington, J.: A transformation system for developing recursive programs. J. ACM **24**(1), 44–67 (1977)
9. Guerrero, J.A., Martínez, M.S., Moreno, G., Vázquez, C.: Designing lattices of truth degrees for fuzzy logic programming environments. In: IEEE Symposium Series on Computational Intelligence, SSCI 2015, Cape Town, South Africa, December 7–10, 2015, pp. 995–1004 IEEE, 2015
10. Guerrero, J.A., Moreno, G.: Optimizing fuzzy logic programs by unfolding, aggregation and folding. Electron. Notes Theor. Comput. Sci. **219**, 19–34 (2008)
11. Hanus, M. (ed.): Curry: An integrated functional logic language (vers. 0.8.3). Available at http://www.curry-language.org, (2012)
12. Huet, G., Lévy, J.J.: Computations in orthogonal rewriting systems, Part I + II. In: Lassez, J.L., Plotkin, G.D. (eds.) Computational logic—essays in honor of alan robinson, pp. 395–443. The MIT Press, Cambridge, MA (1992)
13. Hutton, Graham: Programming in Haskell. Cambridge University Press, Cambridge (2007)

14. Ishizuka, M., Kanai, N.: Prolog-ELF Incorporating Fuzzy Logic. In Joshi, A.K. (eds.) Proceedings of the 9th International Joint Conference on Artificial Intelligence, IJCAI'85, 701–703. Morgan Kaufmann, 1985
15. Julián Iranzo, P., Moreno, G., Penabad, J., Vázquez, C.: A declarative semantics for a fuzzy logic language managing similarities and truth degrees. In: Alferes, J., Bertossi, L.E., Governatori, G., Fodor, P., Roman, D. (eds.) Rule Technologies. Research, Tools, and Applications—10th International Symposium, RuleML 2016, Stony Brook, NY, USA, July 6–9, 2016. Proceedings, volume 9718 of Lecture Notes in Computer Science, 68–82. Springer, 2016
16. Julián, P., Medina, J., Morcillo, P.J., Moreno, G., Ojeda-Aciego, M.: An unfolding-based preprocess for reinforcing thresholds in fuzzy tabulation. In: Proceedings of the 12th International Work-Conference on Artificial Neural Networks, IWANN'13, 647–655. Springer Verlag, LNCS 7902, Part I, 2013
17. Julián, P., Moreno, G., Penabad, J.: On fuzzy unfolding. a multi-adjoint approach. Fuzzy Sets Sys. **154**, 16–33 (2005)
18. Julián, P., Moreno, G., Penabad, J.: Operational/interpretive unfolding of multi-adjoint logic programs. J. Universal Comput. Sci. **12**(11), 1679–1699 (2006)
19. Kifer, M., Subrahmanian, V.S.: Theory of generalized annotated logic programming and its applications. J. Logic Program. **12**, 335–367 (1992)
20. Klop, J.W., Middeldorp, A.: Sequentiality in Orthogonal Term Rewriting Systems. J. Symbol. Comput. 161–195 (1991)
21. Lee, R.C.T.: Fuzzy logic and the resolution principle. J. ACM **19**(1), 119–129 (1972)
22. Lee, C., Jones, N., Ben-Amram, A.: The size-change principle for program termination. SIGPLAN Not. **36**(3), 81–92 (2001)
23. Li, D., Liu, D.: A fuzzy Prolog database system. Wiley, New Yark (1990)
24. Lloyd, J.W.: Foundations of Logic Programming. Springer-Verlag, Berlin (1987)
25. Medina, J., Ojeda-Aciego, M., Vojtáš, P.: Multi-adjoint logic programming with continuous semantics. In: Proc of Logic Programming and Non-Monotonic Reasoning, LPNMR'01, vol. 2173, pp. 351–364. Springer-Verlag, LNAI, (2001)
26. Medina, J., Ojeda-Aciego, M., Vojtáš P.: A procedural semantics for multi-adjoint logic programing. In: Progress in Artificial Intelligence, EPIA'01, vol. 2258(1), pp. 290–297 Springer-Verlag, LNAI, (2001)
27. Medina, J., Ojeda-Aciego, M., Vojtáš, P.: Similarity-based Unification: a multi-adjoint approach. Fuzzy Sets Sys. **146**, 43–62 (2004)
28. Morcillo, P.J., Moreno, G., Penabad, J., Vázquez, C.: Fuzzy Computed Answers Collecting Proof Information. In: Cabestany, J., et al. (eds.) Advances in Computational Intelligence—Proceedings of the 11th International Work-Conference on Artificial Neural Networks, IWANN 2011, 445–452. Springer Verlag, LNCS 6692, 2011
29. Morcillo, P.J., Moreno, G.: Improving multi-adjoint logic programs by unfolding fuzzy connective definitions. In: Proceedings of the 13th International Work-Conference on Artificial Neural Networks, IWANN 2015 Mallorca, Spain, June 10–12, pp. 511–524. Springer Verlag, LNCS 9094, Part I, 2015
30. Morcillo, P.J., Moreno, G.: On cost estimations for executing fuzzy logic programs. In Arabnia, H.R. et al. (ed.) Proceedings of the 11th International Conference on Artificial Intelligence, ICAI 2009, July 13–16, 2009, Las Vegas (Nevada), USA
31. Morcillo, P.J., Moreno, G. Penabad, J., Vázquez, C.: A Practical Management of Fuzzy Truth Degrees using FLOPER. In: Dean, M., et al. (eds.) Proceedings of 4nd Intl Symposium on Rule Interchange and Applications, RuleML'10. Washington, USA, October 21–23, 20–34. Springer Verlag, LNCS 6403, 2010
32. Morcillo, P.J. Moreno, G.: Modeling interpretive steps in fuzzy logic computations. In: Di Gesù V. et al. (eds) Proceedings of the 8th International Workshop on Fuzzy Logic and Applications, WILF 2009. Palermo, Italy, June 9-12, 44–51. Springer Verlag, LNAI 5571, 2009
33. Moreno, G., Penabad, J., Vidal. G.: Tuning fuzzy logic programs with symbolic execution. CoRR, abs/1608.04688, 2016

34. Moreno, G.: Building a fuzzy transformation system. In: Wiedermann, J. et al. (eds.) Proceedings of the 32nd Conference on Current Trends in Theory and Practice of Computer Science, SOFSEM'2006. Merin, Czech Republic, January 21–27, 409–418. Springer Verlag, LNCS 3831, 2006

35. Moreno, Ginés: A narrowing-based instantiation rule for rewriting-based fold/unfold transformations. Electr. Notes Theor. Comput. Sci. **86**(3), 144–167 (2003)

36. Moreno, G., Vázquez, C.: Fuzzy logic programming in action with floper. J. Software Eng. Appl. **7**, 273–298 (2014)

37. Rodríguez-Artalejo, M., Romero-Díaz, C.: Quantitative logic programming revisited. In: Proceedings of 9th Functional and Logic Programming Symposium, FLOPS'08, 272–288. LNCS 4989, Springer Verlag, 2008

38. Straccia, U.: Managing uncertainty and vagueness in description logics, logic programs and description logic programs. In: Reasoning Web, 4th International Summer School, Tutorial Lectures, number 5224 in Lecture Notes in Computer Science, 54–103. Springer Verlag, 2008

39. Tamaki, H., Sato, T.: Unfold/Fold Transformations of Logic Programs. In: Tärnlund, S. (ed.) *Proceedings of Second International Conference on Logic Programming*, 127–139, 1984

40. Vázquez, C., Moreno, G.. Tomás, L., Tordsson, J.: A cloud scheduler assisted by a fuzzy affinity-aware engine. In: Yazici, A., Pal, N.R., Kaymak, U., Martin, T., Ishibuchi, H., Lin, C.-T., Sousa, J.M.C., Tütmez, B. (eds.) 2015 IEEE International Conference on Fuzzy Systems, FUZZ-IEEE 2015, Istanbul, Turkey, August 2–5, 2015, 1–8. IEEE, 2015

On Fuzzy Generalizations of Concept Lattices

Lubomir Antoni, Stanislav Krajči and Ondrej Krídlo

Abstract We provide an overview of different generalizations of formal concept analysis based on Fuzzy logic. Firstly, we recall a common platform for early fuzzy approaches and then, we deal with the data heterogeneity and its various extensions. A second-order formal context makes a bridge between the early fuzzy extensions and the heterogeneous frameworks. A second-order formal context is based on the bonds in L-fuzzy generalization. We present the connections between the related approaches.

Keywords Formal concept analysis · Bonds · External formal contexts

1 Introduction

The study of structures and mappings which allow to analyze the data in various forms is a challenging task. In this way, the first attempts to interpret the lattice theory as concretely as possible and to promote the better communication between lattice theorists and potential users of lattice theory represent the inception for data analysis taking into account the binary relations on the objects and attributes sets [94]. Since the concept hierarchies play an important role here, the term of formal concept analysis has been adopted for this reasoning. Briefly, formal concept analysis scrutinizes an object-attribute block of relational data in bivalent form and the complex foundations were built in [45].

L. Antoni (✉) · S. Krajči · O. Krídlo
Institute of Computer Science, Faculty of Science, Pavol Jozef Šafárik
University in Košice, Jesenná 5, 040 01 Košice, Slovakia
e-mail: lubomir.antoni@upjs.sk

S. Krajči
e-mail: stanislav.krajci@upjs.sk

O. Krídlo
e-mail: ondrej.kridlo@upjs.sk

© Springer International Publishing AG 2018
L. T. Kóczy and J. Medina (eds.), *Interactions Between Computational
Intelligence and Mathematics*, Studies in Computational Intelligence 758,
https://doi.org/10.1007/978-3-319-74681-4_6

For the inception of formal concept analysis, very influential texts arise in [8, 9]. The efficient selection of relevant formal concepts is an interesting and important issue for investigation and several studies have focused on this scalability question in formal concept analysis. The stability index of Kuznetsov [67] represents the proportion of subsets of attributes of a given concept whose closure is equal to the extent of this concept (in an extensional formulation). A high stability index signalizes that extent does not disappear if the extent of some of its attributes is modified. It helps to isolate concepts that appear because of noisy objects in [52]. The complete restoring of the original concept lattice is achieved by stability index in combination with two other indices. The phenomenon of the basic level of concepts is advocated to select important formal concepts in Bělohlávek et al. [22]. Five quantitative approaches on the basic level of concepts and their metrics are comparatively analyzed in [23]. The approaches on selecting of the formal concepts and simplifying the concept lattices are examined by [18, 35, 40, 48–50, 66, 68], as well. A large survey on models and techniques in knowledge processing based on formal concept analysis is given by [84].

The applications of formal concept analysis which focus on the text retrieval and text mining were published in the monograph [34]. The state of art of formal concept analysis and its applications in linguistics, lexical databases, rule mining (iceberg concept lattice), knowledge management, software engineering, object-engineering, software development and data analysis are covered in the special LNAI 3626 Volume devoted to Formal concept analysis—Foundations and Applications and edited by Berhard Ganter, Gerd Stumme and Rudolf Wille in 2005. An extensive overview of applications of formal concept analysis in the various application domains including software mining, web analytics, medicine, biology and chemistry data is given by [83]. In addition, we mention the analysis how students learn to program [87], the techniques for analyzing and improving integrated care pathways [82], evaluation of questionnaires [20], or the morphological image and signal processing [2]. Our applications aim at finding a set of representative symptoms for the disease [51], clustering in a social network area [61] or exploring the educational tasks and objectives of teachers who give the lessons of programming fundamentals [3]. We have also applied formal concept analysis to explore the elements of MIT App Inventor 2 programming language [3].

Up to date, we remind that some other extensions called relational concept analysis, logical concept analysis, triadic concept analysis, temporal concept analysis, rough formal concept analysis and pattern structures are sound and available for thorough study and their application potential.

Conceptual scaling [45] and pattern structures [44] offer the possibility to process the many-valued formal contexts. In this many-valued direction, the truth degrees from Fuzzy logic were advocated by many researchers in effort to promote the representation and interpretation of data in many-valued form. In this paper, we recall early fuzzy extensions in formal concept analysis (Sect. 2), then we stress the representation of fuzzy conjunctions (Sect. 3) in the various generalizations. We start with the common generalized platform (Sect. 3.1), several novel heterogeneous approaches (Sect. 3.2) and formal context of higher order (Sect. 3.3). The last one is based on the external formal contexts, harnessing the early fuzzy extensions to the novel frameworks, and allows to explore the relationships between the formal contexts by using the bonds in their L-fuzzy generalization.

2 Fuzzy Formal Context

The statements that people use to communicate facts about the world are usually not bivalent. The truth of such statements is a matter of degree, rather than being only true or false. Fuzzy logic and fuzzy set theory are frameworks which extend formal concept analysis in various independent ways. Here, we recall the basic definitions of fuzzy formal context. The structures of partially ordered set, complete lattice or residuated lattice are applied here to represent data.

Definition 1 Consider two nonempty sets A and B, an ordered set of truth degrees T and a mapping R such that $R : A \times B \longrightarrow T$. Then, the triple $\langle A, B, R \rangle$ is called a (T)-*fuzzy formal context*. The elements of the sets A and B are called attributes and objects, respectively. The mapping R is called a (T)-*fuzzy incidence relation*.

Definition 2 A residuated lattice is an algebra $\langle L, \wedge, \vee, \otimes, \rightarrow, 0, 1 \rangle$ where 0 is the least element and 1 is the greatest element; $\otimes$ is associative, commutative, and the identity $x \otimes 1$ holds; and adjointness property, i.e. $x \leq y \rightarrow z$ if and only if $x \otimes y \leq z$ holds for each $x, y, z \in L$. A residuated lattice is called complete if $\langle L, \wedge, \vee, 0, 1 \rangle$ is a complete lattice. The operations $\otimes$ and $\rightarrow$ are called fuzzy conjunction and residuum, respectively.

In the definition of (T)-fuzzy formal context, we often take the interval $T = [0, 1]$, because it is a frequent scale of truth degrees in many applications. For such replacement, the shortened notion of fuzzy formal context has been adopted. Analogously, we can consider L-fuzzy formal context, or P-fuzzy formal context, having replaced the interval $[0, 1]$ by the algebraic structures of complete residuated lattice L, partially ordered set P or other plausible scale of truth degrees.

Definition 3 Let A be a nonempty set and let T be an ordered set of truth degrees. The mapping $f : A \longrightarrow T$ is called a fuzzy membership function (also called a T-fuzzy set or a fuzzy set). We denote a set of all fuzzy membership functions $A \longrightarrow T$ by T^A.

The list of early fuzzy approaches is described in Table 1. It summarizes the names of approaches, types of fuzzy formal contexts and encloses references on related extensions.

The proper (symmetric) generalization of the classical concept-forming operators (called also derivation operators or polars, see Ganter and Wille [45]) was introduced by Bělohlávek [11].

Table 1 Early fuzzy approaches

Name of approach	Formal context	Authors
L-fuzzy extension	L-fuzzy	Burusco et al. [28], Pollandt [85], Bělohlávek [11]
One-sided extension	Fuzzy	Krajči [56], Yahia et al. [27]
α-cuts	Fuzzy	Snášel, Vojtáš et al. [91]
Crisply generated ext.	L-fuzzy	Bělohlávek, Sklenář, Zacpal [21]

Definition 4 Let $\langle L, \vee, \wedge, \otimes, \rightarrow, 0, 1 \rangle$ be a complete residuated lattice and let $\langle A, B, R \rangle$ be an L-fuzzy formal context. The mappings $\nearrow: L^B \longrightarrow L^A$ and $\swarrow: L^A \longrightarrow L^B$ given by

$$(\nearrow (g))(a) = g^{\nearrow}(a) = \bigwedge_{b \in B}(g(b) \rightarrow R(a, b)), \tag{1}$$

and

$$(\swarrow (f))(b) = f^{\swarrow}(b) = \bigwedge_{a \in A}(f(a) \rightarrow R(a, b)). \tag{2}$$

are called L-fuzzy concept-forming operators (or L-fuzzy derivation operators, shortened L-fuzzy polars).

The set of pairs

$$L\text{-FCL}(A, B, R) = \{\langle f, g \rangle : f^{\swarrow} = g, g^{\nearrow} = f\} \tag{3}$$

is called the set of all L-fuzzy formal concepts of $\langle A, B, R \rangle$. The elements of the pairs $\langle f, g \rangle \in L\text{-FCL}(A, B, R)$ are called the extents and intents of L-fuzzy formal concepts, respectively. The set of all extents is denoted by $\text{Ext}(A, B, R)$, the set of all intents by $\text{Int}(A, B, R)$.

Bělohlávek [11] proved that L-fuzzy concept-forming operators form an antitone Galois connection. In order theory [39, 45, 78], an antitone Galois connection is given by two opposite mappings between ordered sets. These mappings are order-inverting and a composition of these two mappings yields two closure operators.

Definition 5 An antitone Galois connection between two ordered sets $(P, \leq_P)$ and $(Q, \leq_Q)$ is a pair $\langle g, f \rangle$ of mappings $g : P \rightarrow Q, f : Q \rightarrow P$ such that for all $p \in P$ and $q \in Q$

$$g(p) \leq_Q q \quad if and only if \quad p \geq_P f(q).$$

Other useful properties of L-fuzzy formal concepts have been intensively investigated since their first inception in [11]. The structure of L-fuzzy concept lattice is given by the set of all L-fuzzy formal concepts equipped with a crisp order here. In the research line, the papers [12–14] are oriented to fuzzy Galois connections and L-fuzzy concept lattices which are enhanced by a fuzzy order of L-fuzzy formal concepts. More extensive results about these issues have been proliferated in [15, 17, 19, 24–26] and the more thorough findings can be found in the recently published literature.

2.1 One-Sided Concept Lattices

The extents and intents of L-fuzzy formal concepts are both represented by fuzzy membership functions. In [11], Bělohlávek explains the examples of empirically empty fuzzy extents, partially empty fuzzy extents (which can be interpreted apparently natural) and full crisp extents in effort to interpret L-fuzzy concepts given by planets and their attributes.

Full crisp extents with the maintenance of fuzzy intents became the origin of our approach which was elaborated theoretically [56], as well as practically in our special oriented applications for management of teams [60, 61].

Definition 6 Let $X \subseteq B$ and $\uparrow: \mathcal{P}(B) \longrightarrow [0, 1]^A$. Then, $\uparrow$ is a mapping that assigns to every crisp set X of objects a fuzzy membership function $X^\uparrow$ of attributes, such that a value in a point $a \in A$ is:

$$X^\uparrow(a) = \bigwedge_{b \in X} R(a, b). \qquad (4)$$

Definition 7 Let $f : A \to [0, 1]$ and $\downarrow: [0, 1]^A \longrightarrow \mathcal{P}(B)$. Then, $\downarrow$ is a mapping that assigns to each fuzzy membership function f of attributes the crisp set

$$f^\downarrow = \{b \in B : (\forall a \in A) R(a, b) \geq f(a)\}. \qquad (5)$$

The mappings $\uparrow$ and $\downarrow$ are called the one-sided concept-forming operators. In addition, the composition of (4) and (5) allows us to define the notion of one-sided fuzzy concept.

Definition 8 Let $X \subseteq B$ and $f \in [0, 1]^A$. The pair $\langle X, f \rangle$ is called a one-sided fuzzy concept, if $X^\uparrow = f$ and $f^\downarrow = X$. The crisp set of objects X is called the extent and the fuzzy membership function $X^\uparrow$ is called the intent of one-sided fuzzy concept.

The set of all one-sided fuzzy concepts ordered by inclusion of extents forms a complete lattice, called one-sided fuzzy concept lattice, as introduced in [56]. Moreover, in [56] we proved that this construction is a generalization of classical approach from [45]. The one-sided fuzzy concept lattices for a fuzzy formal context from Fig. 1 and for P-fuzzy formal context from Fig. 2 are illustrated in Figs. 3 and 4.

The selection of relevant concepts from the set of all one-sided fuzzy concepts is an important issue. In this direction, the method of subset quality measure [91] takes into account the idea of α-cuts from Fuzzy logic for searching of significant clusters. The modified Rice-Siff algorithm provides the efficient generation of clusters and gathers

Fig. 1 Example of fuzzy
formal context

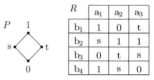

R	a_1	a_2	a_3
b_1	1	0.9	0.8
b_2	0.8	0.7	0.7
b_3	0.3	0.3	0.3
b_4	0.8	0.6	0.9

Fig. 2 Example of P-fuzzy
formal context

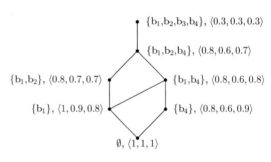

R	a_1	a_2	a_3
b_1	1	0	t
b_2	s	1	1
b_3	0	t	s
b_4	1	s	0

Fig. 3 One-sided fuzzy
concept lattice for formal
context from Fig. 1

$\{b_1,b_2,b_3,b_4\}, \langle 0.3, 0.3, 0.3\rangle$

$\{b_1,b_2,b_4\}, \langle 0.8, 0.6, 0.7\rangle$

$\{b_1,b_2\}, \langle 0.8, 0.7, 0.7\rangle$ $\{b_1,b_4\}, \langle 0.8, 0.6, 0.8\rangle$

$\{b_1\}, \langle 1, 0.9, 0.8\rangle$ $\{b_4\}, \langle 0.8, 0.6, 0.9\rangle$

$\emptyset, \langle 1, 1, 1\rangle$

Fig. 4 One-sided fuzzy
concept lattice for formal
context from Fig. 2

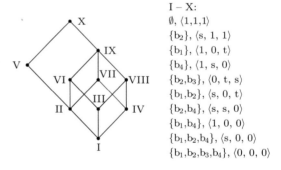

I – X:

$\emptyset, \langle 1,1,1\rangle$
$\{b_2\}, \langle s, 1, 1\rangle$
$\{b_1\}, \langle 1, 0, t\rangle$
$\{b_4\}, \langle 1, s, 0\rangle$
$\{b_2,b_3\}, \langle 0, t, s\rangle$
$\{b_1,b_2\}, \langle s, 0, t\rangle$
$\{b_2,b_4\}, \langle s, s, 0\rangle$
$\{b_1,b_4\}, \langle 1, 0, 0\rangle$
$\{b_1,b_2,b_4\}, \langle s, 0, 0\rangle$
$\{b_1,b_2,b_3,b_4\}, \langle 0, 0, 0\rangle$

the clusters of objects due to their minimal distance function which is specially
defined. The metric properties of this distance function are proved in [56] and the
efficient algorithm is presented and illustrated. Recently, the Gaussian probabilistic
index [5] was introduced in effort to describe the stability of the extents due to random
fluctuation of values in [0, 1]-fuzzy formal context.

The applications of one-sided approach in a social network area, specially in a
secondary school class in Slovakia [60, 61] help to understand the structure of the
managed team or real school classes. The school data are collected persistently and
the mentioned methods for selection of relevant clusters are applied to serve the
complex information for the class teacher about relationships between schoolmates.

Last but not least, the fuzzy subsets in one coordinate and the crisp subsets in the other coordinate of a formal concept are independently investigated in [21, 27], as well. Authors of [21] accentuate the crisply generated fuzzy concepts in effort to deal with the problem of a possibly large number of concepts in an L-fuzzy approach.

3 Representation of Fuzzy Conjunctions and Their Diversification

The relationships between the objects and attributes are represented by a fuzzy incidence relation. Fuzzy conjunctions describe the relationships between the truth degrees of objects and the truth degrees of attributes. In L-fuzzy approach, the fuzzy conjunctions $L \times L \to L$ usually describe these relationships.

The inevitability to represent a diverse form of fuzzy conjunctions in a fuzzy formal context leads to the novel definitions of the fuzzy formal contexts in a broader perspective. Therefore, the notions of generalized formal context, multi-adjoint formal context, heterogeneous formal context, connectional formal context and formal context of second-order were recently introduced. Each of them allows to represent data in a specific form and offers the interpretation on the examples from the real world. The list of novel fuzzy approaches follows in Table 2.

The understanding the way how the fuzzy conjunctions are represented is an important issue for potential applications. In the following subsections, we present the cornerstone of generalized, heterogeneous and second-order version of a formal context.

3.1 Common Generalized Platform

As we have mentioned previously, the fuzzy conjunctions in a fuzzy formal context express the way in which the degrees of objects and degrees of attributes are mutually

Table 2 Novel fuzzy approaches

Name of approach	Formal context	Authors
Generalized extension	Generalized	Krajči [57, 58]
Multi-adjoint extension	Multi-adjoint	Medina and Ojeda-Aciego [69, 71, 74, 75]
Heterogeneous extension	Heterogeneous	Krajči et al. [7], Medina et al. [70], Pócs [79, 80], Popescu [86]
Heterog. one-sided ext.	Heterogeneous one-sided	Butka and Pócs [29, 32]
Higher order extension	Second-order	Krídlo et al. [62]

L \ L	0	0.5	1
0	0	0	0
0.5	0	0.5	0.5
1	0	0.5	1

L \ L	0	0.5	1
0	0	0	0
0.5	0	0	0.5
1	0	0.5	1

L \ L	0	0.5	1
0	0	0	0
0.5	0	0.25	0.5
1	0	0.5	1

Fig. 5 Gödel, Łukasiewicz and product fuzzy conjunctions

connected. In L-fuzzy approach, the frequently used formulas for fuzzy conjunction $\otimes : L \times L \longrightarrow L$ are:

- Gödel fuzzy conjunction given by $a \otimes b = \min(a, b)$,
- Łukasiewicz fuzzy conjunction given by $a \otimes b = \max(0, a + b - 1)$,
- product fuzzy conjunction given by $a \otimes b = a \cdot b$.

The Gödel fuzzy conjunction points to a smaller number of formal concepts in comparison with the Łukasiewicz fuzzy conjunction. Moreover, a smaller fluctuation of fuzzy membership function usually causes a smaller fluctuation of its closure computed by the Łukasiewicz fuzzy conjunction in comparison with the Gödel structure. The product fuzzy conjunction can be discretized to compute the formal concepts in L-fuzzy approach. However, notice that the Gödel, Łukasiewicz and product fuzzy conjunctions are commutative, monotone and left-continuous operations. The examples of Gödel, Łukasiewicz and product fuzzy conjunctions are shown in Fig. 5.

The first prospects to non-commutativity of fuzzy conjunctions $L \times L \longrightarrow L$ is given by Georgescu and Popescu in [47]. This non-commutative extension requires the definition of L-fuzzy concept with one extent and two intents.

Analogously in [58], our motivation is based on a more generalized formulation of fuzzy conjunctions in comparison with L-fuzzy formal context (also for non-commutative cases). The aim is to formulate fuzzy conjunctions by three-sorted structures of truth degrees which are not necessarily numerical, see for instance ($\{s, t, u\}, s \leq t \leq u$) or ($\{\bot, \top\}, \bot \leq \top$). Another possibility is to consider not linearly ordered set of truth degrees, e. g. ($\{0, s, t, 1\}, 0 \leq s \leq 1, 0 \leq t \leq 1$, s is incomparable with t). We aim at preserving the one extent and one intent in a formal concept.

Definition 9 Let $\langle A, B, R \rangle$ be a P-fuzzy formal context for a poset $(P, \preceq_P)$. Let $(C, \preceq_C)$ and $(D, \preceq_D)$ be the complete lattices.

Let $\bullet$ be an operation (a fuzzy conjunction) such that $\bullet$ is from $C \times D$ to P and is monotone and left-continuous in both arguments, that is:

(1a)　$c_1 \preceq_C c_2$ implies $c_1 \bullet d \preceq_P c_2 \bullet d$ for all $c_1, c_2 \in C$ and $d \in D$.

(1b)　$d_1 \preceq_D d_2$ implies $c \bullet d_1 \preceq_P c \bullet d_2$ for all $c \in C$ and $d_1, d_2 \in D$.

(2a)　If $c \bullet d \preceq_P p$ for $d \in D, p \in P$ and for all $c \in S \subseteq C$, then $\bigvee S \bullet d \preceq_P p$.

(2b)　If $c \bullet d \preceq_P p$ for $c \in C, p \in P$ and for all $d \in T \subseteq D$, then $c \bullet \bigvee T \preceq_P p$.

Then, the tuple $\langle A, B, P, R, C, D, \bullet \rangle$ is called a *generalized formal context*.

Fig. 6 General form of two fuzzy conjunctions

D	C 0	0.5	1
0	0	0	0
0.5	0	0	0
1	0	0.5	1

D	C s	t	u
0	⊥	⊥	⊥
0.5	⊥	⊤	⊤
1	⊥	⊤	⊤

The left part of Fig. 6 indicates non-commutative fuzzy conjunction. The right part expresses a fuzzy conjunction in a generalized formal context.

Definition 10 Let $\langle A, B, P, R, C, D, \bullet \rangle$ be a generalized formal context. Let $g : B \longrightarrow D$ and $f : A \longrightarrow C$. Then, the mappings $\nearrow_G : D^B \longrightarrow C^A$ and $\swarrow_G : C^A \longrightarrow D^B$ given by

$$(\nearrow_G(g))(a) = \bigvee \{c \in C : (\forall b \in B) c \bullet g(b) \preceq_P R(a, b)\},$$

$$(\swarrow_G(f))(b) = \bigvee \{d \in D : (\forall a \in A) f(a) \bullet d \preceq_P R(a, b)\}$$

are called the generalized concept-forming operators.

In [57, 58], we proved that a generalized approach is a generalization of L-fuzzy extension from Bělohlávek [14]. It is sufficient to translate the notions of L-fuzzy approach to our generalized approach (see Table 3). In the generalized concept-forming operators, we need to replace a complete lattice $(C, \preceq_C)$, a complete lattice $(D, \preceq_D)$ and a poset $(P, \preceq_P)$ by the complete residuated lattices $\langle L, \vee, \wedge, \otimes, \rightarrow, 0, 1 \rangle$. Then, we can replace the operation $\bullet$ by $\otimes$. In this setting, it can be proved that generalized concept-forming operators and L-fuzzy concept-forming operators coincide (for more details see [58]).

In Fuzzy logic, the hedges are unary connectives which intensify the meaning of the statements. Bělohlávek and Vychodil [25] applied the hedges for the L-fuzzy approach to demonstrate the reduction in size of the L-fuzzy concept lattice. By

Table 3 Translation of L-fuzzy approach to generalized approach

Generalized approach [57, 58]	L-fuzzy approach [14]
Poset $(P, \preceq_P)$	Complete residuated lattice
	$\langle L, \vee, \wedge, \otimes, \rightarrow, 0, 1 \rangle$
Complete lattice $(C, \preceq_C)$	Complete residuated lattice
	$\langle L, \vee, \wedge, \otimes, \rightarrow, 0, 1 \rangle$
Complete lattice $(D, \preceq_D)$	Complete residuated lattice
	$\langle L, \vee, \wedge, \otimes, \rightarrow, 0, 1 \rangle$
$R : A \times B \to P$	$R : A \times B \to L$
$\bullet : C \times D \to P$	$\otimes : L \times L \to L$

experiments, they claim that a stronger hedge lead to a smaller L-fuzzy concept lattice. The close relationship between the generalized concept lattices and L-fuzzy concept lattices constrained by truth-stressing hedges is explored in [59]. The foundations of isotone fuzzy Galois connections with a truth-stressing hedge and a truth-depressing hedge were developed by Konečný [53] and the reduction of trilattices with truth-stressing hedges is described by Konečný and Osička in [55].

3.2 Heterogeneous Approaches

The reasoning about heterogeneous data in a formal context appears in [36–38, 70, 72, 79, 86]. These works inspired us to define the heterogeneous extension based on the full diversification of structures, whereby the construction of concept lattices is still sound and the proposed extension covers the well-known approaches.

Particularly, the main idea is based on a diversification of all data structures that can be diversified within a formal context. We use different sets of truth degrees across a set of objects, different sets of truth degrees across a set of attributes and different sets of truth degrees across the object-attribute table fields. In addition, for each object-attribute pair, one can formulate the original fuzzy conjunctions according to the three-sorted sets of truth degrees of each particular object, attribute and type of values of corresponding table field. The full definition of heterogeneous formal context follows.

Definition 11 Consider the set of attributes A and the set of objects B. Let $\mathcal{P} = ((P_{a,b}, \preceq_{P_{a,b}}) : a \in A, b \in B)$ be a system of posets and let R be a function from $A \times B$ such that $R(a, b) \in P_{a,b}$ for all $a \in A, b \in B$. Let $\mathcal{C} = ((C_a, \preceq_{C_a}) : a \in A)$ and $\mathcal{D} = ((D_b, \preceq_{D_b}) : b \in B)$ be systems of complete lattices.

Let $\odot = (\bullet_{a,b} : a \in A, b \in B)$ be a system of operations such that $\bullet_{a,b}$ is from $C_a \times D_b$ to $P_{a,b}$ and is monotone and left-continuous in both arguments, that is:

(1a) $c_1 \preceq c_2$ implies $c_1 \bullet_{a,b} d \preceq c_2 \bullet_{a,b} d$ for all $c_1, c_2 \in C_a$ and $d \in D_b$.

(1b) $d_1 \preceq d_2$ implies $c \bullet_{a,b} d_1 \preceq c \bullet_{a,b} d_2$ for all $c \in C_a$ and $d_1, d_2 \in D_b$.

(2a)   If $c \bullet_{a,b} d \preceq p$ for $d \in D_b$, $p \in P_{a,b}$ and for all $c \in S \subseteq C_a$, then $\bigvee S \bullet_{a,b} d \preceq p$.

(2b)   If $c \bullet_{a,b} d \preceq p$ for $c \in C_a$, $p \in P_{a,b}$ and for all $d \in T \subseteq D_b$, then $c \bullet_{a,b} \bigvee T \preceq p$.

Then, we call the tuple $\langle A, B, \mathcal{P}, R, \mathcal{C}, \mathcal{D}, \odot \rangle$ a *heterogeneous formal context*.

The direct product of complete lattices $C_1 \times C_2 \times \ldots \times C_n$ can be denoted by $\prod_{i \in \{1,2,\ldots,n\}} C_i$. Then, the heterogeneous concept-forming operators are defined between the direct products of complete lattices.

Definition 12 Let $g \in \prod_{b \in B} D_b$ and $\nearrow_H : \prod_{b \in B} D_b \longrightarrow \prod_{a \in A} C_a$. Then, $\nearrow_H$ is a mapping that assigns to every function g a function $\nearrow_H(g)$, such that its value in the attribute a is given by

D_{b_1}	C_{a_1}		
	s	t	u
0	0	0	0
0.5	0	0.5	0.5
1	0	0.5	1

D_{b_2}	C_{a_1}		
	s	t	u
0	i	i	i
0.33	i	ii	ii
0.66	i	ii	ii
1	i	ii	ii

D_{b_3}	C_{a_1}		
	s	t	u
$\perp$	0	0	0
$\top$	0	1	1

Fig. 7 Heterogeneous form of fuzzy conjunctions

$$(\nearrow_H(g))(a) = \bigvee \{c \in C_a : (\forall b \in B) c \bullet_{a,b} g(b) \preceq R(a,b)\}.$$

Symmetrically, let $f \in \prod_{a \in A} C_a$ and $\swarrow_H : \prod_{a \in A} C_a \longrightarrow \prod_{b \in B} D_b$. Then, $\swarrow_H$ is a mapping that assigns to every function f a function $\swarrow_H(f)$, such that its value in the object b is given by

$$(\swarrow_H(f))(b) = \bigvee \{d \in D_b : (\forall a \in A) f(a) \bullet_{a,b} d \preceq R(a,b)\}.$$

The mappings $\nearrow_H$ and $\swarrow_H$ are called heterogeneous concept-forming operators. A pair $\langle g, f \rangle$ such that $\nearrow_H(g) = f$ and $\swarrow_H(f) = g$ is called a heterogeneous formal concept. The ordering of heterogeneous formal concepts $\langle g_1, f_1 \rangle \leq \langle g_2, f_2 \rangle$ is defined by $g_1 \leq g_2$ (or equivalently $f_1 \geq f_2$).

Then, the poset of all heterogeneous formal concepts ordered by $\leq$ will be called a *heterogeneous concept lattice*. We denote a heterogeneous concept lattice by $\text{HCL}(A, B, \mathcal{P}, R, \mathcal{C}, \mathcal{D}, \odot, \swarrow_H, \nearrow_H, \leq)$.

The fuzzy conjunctions in a heterogeneous form are exemplified in Fig. 7. It represents the fuzzy conjunctions between one particular attribute and three different objects.

Illustration of such heterogeneous formal context on an example of longterm and shortterm preferences of people can be found in [7]. Moreover, the basic theorem on heterogeneous concept lattices is formulated and proved there.

Theorem 1 (Basic theorem on heterogeneous concept lattices)

1. *A heterogeneous concept lattice* $\text{HCL}(A, B, \mathcal{P}, R, \mathcal{C}, \mathcal{D}, \odot, \swarrow_H, \nearrow_H, \leq)$ *is a complete lattice in which*

$$\bigwedge_{i \in I} \langle g_i, f_i \rangle = \left\langle \bigwedge_{i \in I} g_i, \nearrow_H \left(\swarrow_H \left(\bigvee_{i \in I} f_i \right) \right) \right\rangle$$

and

$$\bigvee_{i \in I} \langle g_i, f_i \rangle = \left\langle \swarrow_H \left(\nearrow_H \left(\bigvee_{i \in I} g_i \right) \right), \bigwedge_{i \in I} f_i \right\rangle.$$

2. *For each $a \in A$ and $b \in B$, let $0_{P_{a,b}}$ be the least element of $P_{a,b}$ such that $0_{C_a} \bullet_{a,b} d = c \bullet_{a,b} 0_{D_b} = 0_{P_{a,b}}$ for all $c \in C_a$, $d \in D_b$. Then, a complete lattice L is isomorphic to $HCL(A, B, \mathcal{P}, R, \mathcal{C}, \mathcal{D}, \odot, \swarrow_H, \nearrow_H, \leq)$ if and only if there are mappings $\alpha : \bigcup_{a \in A}(\{a\} \times C_a) \longrightarrow L$ and $\beta : \bigcup_{b \in B}(\{b\} \times D_b) \longrightarrow L$ such that:*

   (1a) α does not increase in the second argument (for a fixed first argument);
   (1b) β does not decrease in the second argument (for a fixed first argument);
   (2a) $Rng(\alpha)$ is $\wedge$-dense in L;
   (2b) $Rng(\beta)$ is $\vee$-dense in L; and
   (3)   For every $a \in A$, $b \in B$, $c \in C_a$ and $d \in D_b$,

$$\alpha(a, c) \geq \beta(b, d) \quad \text{if and only if} \quad c \bullet_{a,b} d \leq R(a, b).$$

Proof The proof is presented in [7]. ∎

Other inspiring approaches describe the possibility to apply formal concept analysis for heterogeneous data, as well. Pócs introduces the connectional approach in [79, 80] which is based on the antitone Galois connections (Definition 5) between each pair of object and attribute. Fuzzy conjunctions are replaced by the antitone Galois connections. Note that the definition of connectional formal context does not contain the fuzzy incidence relation.

Definition 13 Let A and B be non-empty sets. Let $\mathcal{C} = ((C_a, \preceq_{C_a}) : a \in A)$ and $\mathcal{D} = ((D_b, \preceq_{D_b}) : b \in B)$ be systems of complete lattices. Let $\mathcal{G} = ((\phi_{a,b}, \psi_{a,b}) : a \in A, b \in B)$ be a system of antitone Galois connections, such that $(\phi_{a,b}, \psi_{a,b})$ is an antitone Galois connection between $(C_a, \preceq_{C_a})$ and $(D_b, \preceq_{D_b})$. Then, the tuple $\langle A, B, \mathcal{G}, \mathcal{C}, \mathcal{D} \rangle$ is called a *connectional formal context*.

For $g \in \prod_{b \in B} D_b$, the mapping $\nearrow_{\mathcal{G}} : \prod_{b \in B} D_b \longrightarrow \prod_{a \in A} C_a$ is defined by

$$(\nearrow_{\mathcal{G}}(g))(a) = \bigwedge_{b \in B} \phi_{a,b}(g(b)).$$

For $f \in \prod_{a \in A} C_a$, the mapping $\swarrow_{\mathcal{G}} : \prod_{a \in A} C_a \longrightarrow \prod_{b \in B} D_b$ is defined by

$$(\swarrow_{\mathcal{G}}(f))(b) = \bigwedge_{a \in A} \psi_{a,b}(f(a)).$$

The mappings $\nearrow_{\mathcal{G}}$ and $\swarrow_{\mathcal{G}}$ are called connectional concept-forming operators.

The connectional formal concepts can be constructed by the connectional concept-forming operators. An ordered set of connectional formal concepts forms a lattice structure [79, 80]. Relationship between heterogeneous approach and connectional approach is described formally in [6].

Figure 8 illustrates an example of antitone Galois connection between the object-attribute pair (a_1, b_1) for $a_1 \in A$, $b_1 \in B$. It is uniquely determined from Fig. 8 and

Fig. 8 Antitone Galois connection from connectional approach

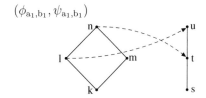

from the properties of an antitone Galois connection that $\phi_{a_1,b_1}(k) = u$, $\phi_{a_1,b_1}(l) = u$, $\phi_{a_1,b_1}(m) = t$, $\phi_{a_1,b_1}(n) = t$, $\psi_{a_1,b_1}(s) = n$, $\psi_{a_1,b_1}(t) = n$ and $\psi_{a_1,b_1}(u) = l$.

Medina and Ojeda-Aciego applied the idea of multi-adjointness from logic programming [76, 77] to formal concept analysis. The multi-adjoint formal context is characterized by the adjoints pertaining to the operations • in a common generalized platform. The first prospects, motivating results and examples were published in [73, 75]. We recall a multi-adjoint formal context here. Two central notions are an adjoint triple and a multi-adjoint frame.

Definition 14 Let $(U, \preceq_U)$, $(V, \preceq_V)$ and $(P, \preceq_P)$ be posets. Then, the triple $(\bullet, \rightarrow_1, \rightarrow_2)$ is called an adjoint triple (or implication triple) if $\bullet : (U \times V) \longrightarrow P$, $\rightarrow_1 : (V \times P) \longrightarrow U$ and $\rightarrow_2 : (U \times P) \longrightarrow V$ and moreover

$$(u \bullet v) \preceq_P p \ \text{ iff } \ u \preceq_U (v \rightarrow_1 p) \ \text{ iff } \ v \preceq_V (u \rightarrow_2 p).$$

Let $(C, \preceq_C)$, $(D, \preceq_D)$ be complete lattices and $(P, \preceq_P)$ be a poset. A multi-adjoint frame is a tuple $\langle C, D, P, \bullet_b : b \in B \rangle$ in which $(\bullet_b, \rightarrow_{1_b}, \rightarrow_{2_b})$ is an adjoint triple with respect to $(C, \preceq_C)$, $(D, \preceq_D)$ and $(P, \preceq_P)$ for all $b \in B$.

Finally, a *multi-adjoint context* is a tuple $\langle A, B, R, \sigma \rangle$ such that A and B are the sets of attributes and objects, respectively, R is a function from $A \times B$ such that $R(a, b) \in P$ for all $a \in A$ and $b \in B$, and σ is a mapping that associates any object $b \in B$ with a particular adjoint triple (shortly denoted by $\sigma(b)$) from the multi-adjoint frame.

For $g \in D^B$, the mapping $\nearrow_\sigma : D^B \longrightarrow C^A$ is defined by

$$(\nearrow_\sigma (g))(a) = \bigwedge_{b \in B} (g(b) \rightarrow_{1_\sigma(b)} R(a, b)).$$

For $f \in C^A$, the mapping $\swarrow_\sigma : C^A \longrightarrow D^B$ is defined by

$$(\swarrow_\sigma (f))(b) = \bigwedge_{a \in A} (f(a) \rightarrow_{2_\sigma(b)} R(a, b)).$$

The mappings $\nearrow_\sigma$ and $\swarrow_\sigma$ are called multi-adjoint concept-forming operators.

The multi-adjoint formal concepts are defined by means of the multi-adjoint concept-forming operators, similarly as in a heterogeneous framework. Moreover, the ordered set of all multi-adjoint formal concepts forms a lattice structure [73, 75].

We remind that the framework of multi-adjoint concept lattices represents the point of interest of many researchers. The extended approaches of t-concept lattices [69], multi-adjoint concept lattices based on heterogeneous conjunctors [70], dual multi-adjoint concept lattices [71], or constraints with hedges [54] are sound and interpretable. The weaker conditions which are sufficient to generate multi-adjoint concept lattices have been recently introduced in [41]. In general, the necessary and sufficient conditions to generate adjunctions [46] and the algebraic properties of multilattices [33] are the point of interest of current research which is closely related to these issues.

The close relationships between the multi-adjoint, heterogeneous and connectional approaches are thoroughly presented in [4, 6, 7] and transformations between them are advocated. Moreover, some other general algebraic aspects of forming fuzzy concept lattices are described in [81]. A biresiduated mapping (acting on two complete lattices and a poset) is defined for every object-attribute pair in a fuzzy formal context. Then, Pócs and Pócsová have proved that for every biresiduated mapping (acting on two complete lattices and a poset), one can construct Galois connection and moreover, the construction of Galois connection works also between direct products of complete lattices, which uniquely determines the corresponding fuzzy concept-forming operators. The theoretical details regarding the representation of the fuzzy concept lattices in the framework of classical concept lattices are described in [30].

The ideas of heterogeneity applied for the one-sided fuzzy approach (see Sect. 2.1) are introduced by Butka and Pócs in [29], whereby the possible applications of heterogeneous one-sided fuzzy concept lattices are advocated in merging of heterogeneous subcontexts.

Definition 15 Consider the set of attributes A and the set of objects B. Let $C = ((C_a, \preceq_{C_a}) : a \in A)$ be a system of complete lattices. Let R be a function such that $R(a, b) \in C_a$ for all $a \in A, b \in B$. Then, we call a tuple $\langle A, B, R, C \rangle$ a *heterogeneous one-sided fuzzy formal context* (or shortly C-*fuzzy formal context*).

By a proper generalization of one-sided fuzzy concept-forming operators from Equation (4) and Equation (5) can be constructed the heterogeneous one-sided fuzzy concept lattices (or shortly C-fuzzy concept lattices). An example of C-fuzzy formal context is shown in Fig. 9.

The relationship between the conceptual scaling and the heterogeneous one-sided fuzzy concept lattices is proved and the method for selecting the relevant heterogeneous one-sided fuzzy formal concepts is presented in [31, 32]. The term of general-

Fig. 9 Example of C-fuzzy formal context

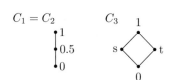

R	a_1	a_2	a_3
b_1	1	0	t
b_2	0.5	1	1
b_3	0	0.5	s
b_4	1	0.5	0

ized one-sided fuzzy approach has been adopted there, but we call it heterogeneous in effort to emphasize the heterogeneous types of attributes in this extension.

An alternative approach to solve the presence of heterogeneous data in a fuzzy formal context, taking into account the structures of idempotent semirings, can be seen in the papers [92, 93] of Valverde-Albacete and Peláez-Moreno.

3.3 Second-Order Formal Context

The main contribution in this subsection is the presentation of novel extended approach in order to merge L-fuzzy framework and heterogeneous framework. The name second-order comes from the idea that one can take an auxiliary concept lattice of some L-fuzzy formal context (entitled as external formal context) and then, the elements of auxiliary concept lattice are taken for the ordered set of truth degrees for some object of the main formal context. This is performed for each object and each attribute of the main formal context. This means that several concept lattices of external formal contexts are used to compute the summary concept lattice.

We define a second order formal context and its derivation operators, moreover present the structural properties of this extension in [62].

Definition 16 Consider two non-empty index sets J, I and L-fuzzy formal context $\langle \bigcup_{j \in J} A_j, \bigcup_{i \in I} B_i, R \rangle$, whereby

- $A_{j_1} \cap A_{j_2} = \emptyset$ for any $j_1, j_2 \in J$, $j_1 \neq j_2$,
- $B_{i_1} \cap B_{i_2} = \emptyset$ for any $i_1, i_2 \in I$, $i_1 \neq i_2$,
- $R : \bigcup_{j \in J} A_j \times \bigcup_{i \in I} B_i \longrightarrow L$.

Moreover, consider two non-empty sets of L-fuzzy formal contexts (external formal contexts) notated by

- $\{ \langle O_j, A_j, Q_j \rangle : j \in J \}$, whereby $\mathcal{D}_j = \langle O_j, A_j, Q_j \rangle$,
- $\{ \langle B_i, T_i, P_i \rangle : i \in I \}$, whereby $\mathcal{C}_i = \langle B_i, T_i, P_i \rangle$.

A *second-order formal context* is a tuple

$$\left\langle \bigcup_{j \in J} A_j, \{\mathcal{D}_j; j \in J\}, \bigcup_{i \in I} B_i, \{\mathcal{C}_i; i \in I\}, R \right\rangle,$$

whereby a subrelation $R_{j,i} : A_j \times B_i \longrightarrow L$ is defined as $R_{j,i}(a, b) = R(a, b)$ for any $a \in A_j$ and $b \in B_i$.

Example of a second-order formal context for two external formal contexts on both objects and attributes sides is shown in Fig. 10.

Particularly, the objects (and attributes) of the second order formal context can be collected by merging the objects (and attributes) of two or more L-fuzzy formal contexts (i. e. external formal contexts). Nevertheless, Definition 16 shows that we

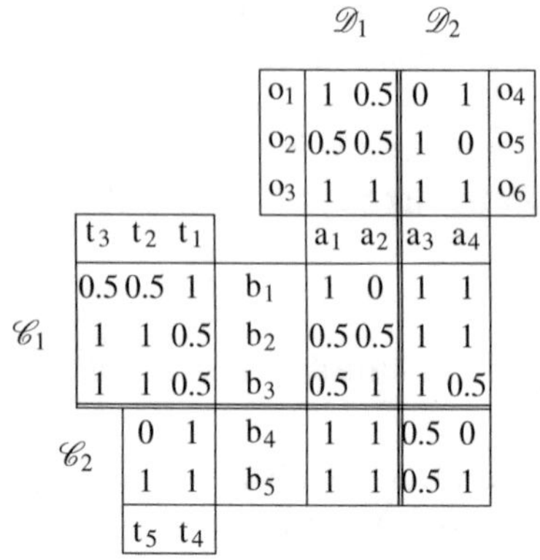

Fig. 10 Example of second-order formal context

$\mathcal{D}_1$ $\mathcal{D}_2$

o_1	1 0.5	0 1	o_4
o_2	0.5 0.5	1 0	o_5
o_3	1 1	1 1	o_6

$t_3\ t_2\ t_1$		$a_1\ a_2$	$a_3\ a_4$
0.5 0.5 1	b_1	1 0	1 1
1 1 0.5	b_2	0.5 0.5	1 1
1 1 0.5	b_3	0.5 1	1 0.5
0 1	b_4	1 1	0.5 0
1 1	b_5	1 1	0.5 1
$t_5\ t_4$			

$\mathcal{C}_1$ (upper rows), $\mathcal{C}_2$ (lower rows)

can also consider only one external formal context for objects and only one external formal context for attributes in a very special case. Furthermore, the extents and intents of these external formal contexts represent the inputs for the computations of second-order formal concepts.

The direct product of external L-fuzzy lattices $L\text{-FCL}(\mathcal{C}_1) \times L\text{-FCL}(\mathcal{C}_2) \times \ldots \times L\text{-FCL}(\mathcal{C}_n)$ is denoted by $\prod_{i \in \{1,2,\ldots,n\}} L\text{-FCL}(\mathcal{C}_i)$. To find the second-order formal concepts, the second-order concept-forming operators are defined between direct products of the two previously described sets of L-fuzzy concept lattices. The precise form of the concept-forming operators are introduced in [62] and their formulation suppose the equivalence functor between the categories of fuzzy Chu correspondences and completely lattice L-ordered sets. For more details, see [62, 65].

Nevertheless, to compute the second-order formal concepts in a more shortened way, we can use the additional results presented in [62]. To do that, we need to recall that given two arbitrary sets A and B, the mapping $\beta : B \to L^A$ is called L-multimapping. Then, a notion of L-bond is defined as follows.

Definition 17 Let $\mathcal{C}_i = \langle A_i, B_i, R_i \rangle$ for $i \in \{1, 2\}$ be two L-fuzzy formal contexts. An L-bond is an L-multimapping $\beta : B_1 \longrightarrow \text{Int}(\mathcal{C}_2)$ such that $\beta^t : A_2 \longrightarrow \text{Ext}(\mathcal{C}_1)$, where $\beta^t(a_2)(b_1) = \beta(b_1)(a_2)$ for any $(b_1, a_2) \in B_1 \times A_2$. The set of all L-bonds between L-fuzzy formal contexts $\mathcal{C}_1$ and $\mathcal{C}_2$ is denoted by $L\text{-Bonds}(\mathcal{C}_1, \mathcal{C}_2)$.

Remark 1 The set of all L-bonds between two L-fuzzy formal contexts forms the structure of complete lattice. For the proof and the precise formulation of supremum and infimum of L-bonds see [64]. Relationships between L-bonds and extents of direct products of L-fuzzy contexts are drawn in [63].

The complete lattice of the nine L-bonds between $\mathcal{D}_1$ and $\mathcal{C}_1$ (from Fig. 10) can be seen in Fig. 11. For the calculations of the extents and intents of L-fuzzy formal contexts by Equation (1) and (2), we selected the Łukasiewicz fuzzy implications given by $a \to b = \min\{1, 1 - a + b\}$ for $a, b \in L$.

Having selected one particular L-bond β from the set of all L-bonds between two L-fuzzy formal contexts, one can introduce the concept-forming operators between

Fig. 11 Concept lattice of
L-bonds between $\mathcal{D}_1$ and $\mathcal{C}_1$
from Fig. 10

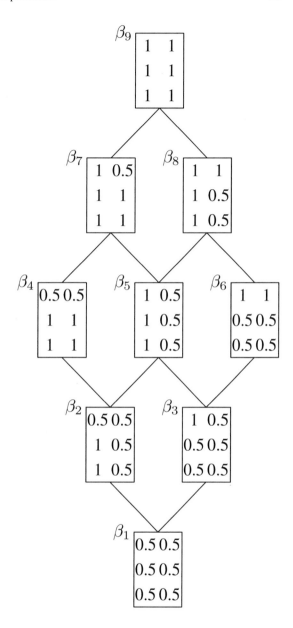

two external sets of all L-fuzzy membership functions (L^{B_1} and L^{A_2}) as the following definition states.

Definition 18 Let $\mathcal{C}_i = \langle A_i, B_i, R_i \rangle$ for $i \in \{1, 2\}$ be two L-fuzzy contexts and let β be an arbitrary L-bond between $\mathcal{C}_1$ and $\mathcal{C}_2$. The mappings $\uparrow_\beta : L^{B_1} \longrightarrow L^{A_2}$ and $\downarrow_\beta : L^{A_2} \longrightarrow L^{B_1}$ such that

$$(\uparrow_\beta (f))(a) = \bigwedge_{b \in B_1} (f(b) \to \beta(b)(a)) \tag{6}$$

and

$$(\downarrow_\beta (g))(b) = \bigwedge_{a \in A_2} (g(a) \to \beta(b)(a)) \tag{7}$$

for any $f \in L^{B_1}$ and $g \in L^{A_2}$ are called the second-order concept-forming operators pertaining the L-bond β .

Remark 2 From [62], we know that a pair $\langle \uparrow_\beta, \downarrow_\beta \rangle$ forms an antitone Galois connection between complete lattices $\langle L\text{-Ext}(\mathcal{C}_1), \leq \rangle$ and $\langle L\text{-Int}(\mathcal{C}_2), \leq \rangle$, where $\leq$ is an ordering based on fuzzy sets inclusion.

It can be proved that the second-order concept lattices and the special L-fuzzy concept lattices are isomorphic as the following theorem states.

Theorem 2 *Let*

$$\mathcal{K} = \left\langle \bigcup_{j \in J} A_j, \{\mathcal{D}_j : j \in J\}, \bigcup_{i \in I} B_i, \{\mathcal{C}_i : i \in I\}, R \right\rangle$$

be a second-order formal context and let

$$\widehat{\mathcal{K}} = \left\langle \bigcup_{j \in J} A_j, \bigcup_{i \in I} B_i, \bigcup_{(j,i) \in J \times I} \rho_{ji} \right\rangle,$$

$$\rho_{ji} = \bigwedge \{\beta \in L\text{-Bonds}(\mathcal{D}_j, \mathcal{C}_i) : (\forall (a_j, b_i) \in A_j \times B_i)\beta(a_j)(b_i) \geq R_{ji}(a_j, b_i)\}$$

be an L-fuzzy formal context. Then, the concept lattices of $\mathcal{K}$ and $\widehat{\mathcal{K}}$ are isomorphic.

Proof The proof is presented in [62].

In an L-fuzzy formal context $\widehat{\mathcal{K}}$ given by Theorem 2, the subrelation R_{ji} of a second-order formal context $\mathcal{K}$ is replaced by an L-bond ρ_{ji} for each pair $(j, i) \in J \times I$. The bond ρ_{ji} is constructed as the closest bond with respect to the subrelation R_{ji}. For more details and examples, see [62]. In the second-order formal context given by Fig. 10, we have that $\rho_{1,1} = \beta_7$, whereby β_7 is described in Fig. 11.

We have described the way how to find the second-order formal concepts. The heterogeneous formal contexts from Definition 11 can be seen in terms of the second-order formal contexts as the following translation indicates:

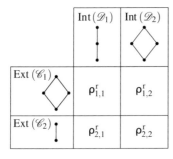
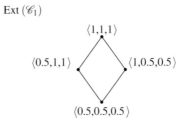

Fig. 12 Heterogeneous formal context derived from second-order formal context

- for attributes and objects take index sets J and I,
- for the sets of truth degrees D_j, C_i for any $(j, i) \in J \times I$ take the complete lattices $\langle \text{Int}(\mathcal{D}_j), \leq \rangle$ and $\langle \text{Ext}(\mathcal{C}_i), \leq \rangle$,
- for the set of truth degrees of table field $P_{j,i}$ take a complete lattice of all fuzzy relations $L^{A_j \times B_i}$ for any $(j, i) \in J \times I$,
- for the values of fuzzy incidence relation R take $R(j, i) = \rho^R_{j,i}$, whereby $\rho^R_{j,i} : A_j \times B_i \longrightarrow L$ such that $\rho^R_{j,i}(a, b) = \rho_{j,i}(a)(b)$; i. e. the closest covering bond $\rho_{j,i}$ is converted to a subrelation $\rho^R_{j,i}$ for any $(j, i) \in J \times I$,
- for operations $\bullet_{j,i} : \text{Int}(\mathcal{D}_j) \times \text{Ext}(\mathcal{C}_i) \longrightarrow L^{A_j \times B_i}$ take

$$(g \bullet_{j,i} f)(a, b) = g(a) \otimes f(b)$$

for any $g \in \text{Int}(\mathcal{D}_j)$ and $f \in \text{Ext}(\mathcal{C}_i)$ and any $(a, b) \in A_j \times B_i$.

Since the operation $\otimes$ (Definition 2) is isotone, then $\bullet_{j,i}$ is isotone for $(j, i) \in J \times I$. In [62], we proved that $\bullet_{j,i}$ is left-continuous for $(j, i) \in J \times I$, hence the assumptions on fuzzy conjunctions from heterogeneous approach [7] are fulfilled.

Heterogeneous formal context in Fig. 12 is derived from the second-order formal context in Fig. 10. We illustrate the exact values of $\text{Ext}(\mathcal{C}_1)$, however the form of $\text{Ext}(\mathcal{C}_i), \text{Int}(\mathcal{D}_j), \rho^R_{j,i}$ for $j, i \in \{1, 2\}$ can be computed from Fig. 10.

4 Conclusion

The information hidden within data can help to solve the many pending issues within community, enterprise or science. Turning of data into knowledge and wisdom is beneficial and necessary, considering either the simple computing in the spreadsheet calculators or various methods of data analysis which are more complex. Data collecting, preprocessing, reduction, visualization and dependencies exploration are important parts of the scientific research, as well.

Fig. 13 Scheme of relationships between approaches

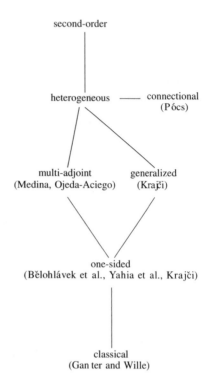

We have presented the various approaches in formal concept analysis—from the early fuzzy approaches to the most recent heterogeneous extensions and the second-order platform. We presented the one-sided fuzzy approach, the common generalized platform, the heterogeneous approaches, the second-order platform and a translation to find the second-order formal concepts. The extensive references on papers dealing with the related generalizations and close areas are included and commented throughout the paper.

We have also recalled the frameworks of multi-adjoint formal context and connectional formal context which offer the base for the various inspiring extensions. For mathematical aspects of relationships between the presented extensions, we refer to the papers [4, 6, 7, 31, 32, 56]. A scheme of these relationships are illustrated in Fig. 13.

An interval-valued fuzzy setting [1] and the bipolar fuzzy membership functions which represent the positive and negative information in L-fuzzy formal contexts [88, 89] seem to be the potential areas for a continuation of a current research line in the heterogeneous extensions. The results of possibility theory [42], probability theory [43, 90] or fuzzy relational equations [10, 16] can be harnessed to characterize the (fuzzy) concept lattices.

In conclusion, we are convinced that the area of concept lattices plays an important role in the research of both theoreticians and practitioners and thus, the research from a broader perspective is still beneficial.

Acknowledgements We thank the anonymous reviewers for their careful reading of our manuscript and their many fruitful comments and suggestions. This work was partially supported by the Scientific Grant Agency of the Ministry of Education of Slovak Republic and the Slovak Academy of Sciences under the contract VEGA 1/0073/15 and by the Slovak Research and Development Agency under the contract No. APVV–15–0091. This work was partially supported by the Agency of the Ministry of Education, Science, Research and Sport of the Slovak Republic for the Structural Funds of EU under the project Center of knowledge and information systems in Košice, CeZIS, ITMS: 26220220158.

References

1. Alcalde, C., Burusco, A., Fuentes-González, R., Zubia, I.: The use of linguistic variables and fuzzy propositions in the L-fuzzy concept theory. Comput. Math. Appl. **62**(8), 3111–3122 (2011)
2. Alcalde, C., Burusco, A., Fuentes-González, R.: Application of the L-fuzzy concept analysis in the morphological image and signal processing. Ann. Math. Artif. Intell. **72**(1–2), 115–128 (2014)
3. Antoni, L., Guniš, J., Krajči, S., Krídlo, O., Šnajder, L.: The educational tasks and objectives system within a formal context. In: Bertet, K., Rudolph, S. (eds.) CLA 2014. CEUR Workshop Proceedings, vol. 1252, pp. 35–46. CEUR-WS.org (2014)
4. Antoni, L., Krajči, S., Krídlo, O.: On different types of heterogeneous formal contexts. In: Pasi, G., Montero, J., Ciucci, D. (eds.) Proceedings of the 8th Conference of the European Society for Fuzzy Logic and Technology, pp. 302–309. Atlantis Press (2013)
5. Antoni, L., Krajči, S., Krídlo, O.: Randomized fuzzy formal contexts and relevance of one-sided concepts. In: Baixeries, J., Sacarea, Ch. (eds.) ICFCA 2015. LNCS (LNAI), vol. 9113. Springer, Heidelberg (2015)
6. Antoni, L., Krajči, S., Krídlo, O., Macek, B., Pisková, L.: Relationship between two FCA approaches on heterogeneous formal contexts. In: Szathmary, L., Priss, U. (eds.) CLA 2012. CEUR Workshop Proceedings, vol. 972, pp. 93–102. CEUR-WS.org (2012)
7. Antoni, L., Krajči, S., Krídlo, O., Macek, B., Pisková, L.: On heterogeneous formal contexts. Fuzzy Sets Syst. **234**, 22–33 (2014)
8. Arnauld, A., Nicole, P.: Logic, or the Art of Thinking. Cambridge University Press, Cambridge (1996)
9. Barbut, M., Monjardet, B.: Ordre et classification. Algbre et combinatoire. Collection Hachette Universit, Librairie Hachette (1970)
10. Bartl, E.: Minimal solutions of generalized fuzzy relational equations: probabilistic algorithm based on greedy approach. Fuzzy Sets Syst. **260**, 25–42 (2015)
11. Bělohlávek, R.: Lattices generated by binary fuzzy relations. Tatra Mt. Math. Publ. **16**, 11–19 (1999)
12. Bělohlávek, R.: Fuzzy galois connections. Math. Logic Q. **45**(4), 497–504 (1999)
13. Bělohlávek, R.: Reduction and a simple proof of characterization of fuzzy concept lattices. Fundam. Informaticae **46**(4), 277–285 (2001)
14. Bělohlávek, R.: Concept lattices and order in fuzzy logic. Ann. Pure Appl. Logic **128**, 277–298 (2004)
15. Bělohlávek, R.: What is fuzzy concept lattice? II. In: Kuznetsov, S.O., Slezak, D., Hepting, D.H., Mirkin, B.G. (eds.) Rough Sets, Fuzzy Sets, Data Mining and Granular Computing. LNCS, vol. 6743, pp. 19–26. Springer, Heidelberg (2011)

16. Bělohlávek, R.: Sup-t-norm and inf-residuum are one type of relational product: Unifying framework and consequences. Fuzzy Sets Syst. **197**, 45–58 (2012)
17. Bělohlávek, R.: Ordinally equivalent data: a measurement-theoretic look at formal concept analysis of fuzzy attributes. Int. J. Approx. Reason. **54**(9), 1496–1506 (2013)
18. Bělohlávek, R., De Baets, B., Konečný, J.: Granularity of attributes in formal concept analysis. Inf. Sci. **260**, 149–170 (2014)
19. Bělohlávek, R., Klir, G.J.: Concepts and Fuzzy Logic. MIT Press, Cambridge (2011)
20. Bělohlávek, R., Sigmund, E., Zacpal, J.: Evaluation of IPAQ questionnaires supported by formal concept analysis. Inf. Sci. **181**(10), 1774–1786 (2011)
21. Bělohlávek, R., Sklenář, V., Zacpal, J.: Crisply generated fuzzy concepts. In: Ganter, B., Godin, R. (eds.) ICFCA 2005. LNCS (LNAI), vol. 3403, pp. 269–284. Springer, Heidelberg (2005)
22. Bělohlávek, R., Trnečka, M.: Basic level of concepts in formal concept analysis. In: Domenach, F., Ignatov, D., Poelmans, J. (eds.) ICFCA 2012. LNCS (LNAI), vol. 7278, pp. 28–44. Springer, Heidelberg (2012)
23. Bělohlávek, R., Trnečka, M.: Basic Level in Formal Concept Analysis: Interesting Concepts and Psychological Ramifications. In: Rossi, F. (eds.) IJCAI 2013, pp. 1233–1239. AAAI Press (2013)
24. Bělohlávek, R., Vychodil, V.: What is fuzzy concept lattice?. In: Bělohlávek, R., Snášel, V. (eds.) CLA 2005. CEUR Workshop Proceedings, vol. 162, pp. 34–45. CEUR-WS.org (2005)
25. Bělohlávek, R., Vychodil, V.: Fuzzy concept lattices constrained by hedges. J. Adv. Comput. Intell. Intell. Inform. **11**(6), 536–545 (2007)
26. Bělohlávek, R., Vychodil, V.: Formal concept analysis and linguistic hedges. Int. J. Gen. Syst. **41**(5), 503–532 (2012)
27. Ben Yahia, S., Jaoua, A.,: Discovering knowledge from fuzzy concept lattice. In: Kandel, A., Last, M., Bunke, H. (eds.) Data Mining and Computational Intelligence. Studies in Fuzziness and Soft Computing, vol. 68, pp. 167–190. Physica-Verlag, Heidelberg (2001)
28. Burusco, A., Fuentes-González, R.: The study of *L*-fuzzy concept lattice. Mathw. Soft Comput. **3**, 209–218 (1994)
29. Butka, P., Pócs, J.: Generalization of one-sided concept lattices. Comput. Inform. **32**(2), 355–370 (2013)
30. Butka, P., Pócs, J., Pócsová, J.: Representation of fuzzy concept lattices in the framework of classical FCA. J. Appl. Math. 2013, Article ID 236725, 7 pages. (2013)
31. Butka, P., Pócs, J., Pócsová, J.: On equivalence of conceptual scaling and generalized one-sided concept lattices. Inf. Sci. **259**, 57–70 (2014)
32. Butka, P., Pócs, J., Pócsová, J.: Reduction of concepts from generalized one-sided concept lattice based on subsets quality measure. In: Zgrzywa, A., Choros, A., Sieminski, A. (eds.) New Research in Multimedia and Internet Systems. Advances in Intelligent Systems and Computing, vol. 314, pp. 101–111. Springer (2015)
33. Cabrera, I.P., Cordero, P., Gutiérez, G., Martinez, J., Ojeda-Aciego, M.: On residuation in multilattices: filters, congruences, and homomorphisms. Fuzzy Sets Syst. **234**, 1–21 (2014)
34. Carpineto, C., Romano, G.: Concept Data Analysis. Wiley, Theory and Applications. J (2004)
35. Cellier, P., Ferré, S., Ridoux, O., Ducassé, M.: A parameterized algorithm to explore formal contexts with a taxonomy. Int. J. Found. of Comput. Sci. **2**, 319–343 (2008)
36. Cornejo, M.E., Medina, J., Ramírez-Poussa, E.: Attribute and size reduction mechanisms in multi-adjoint concept lattices. J. Comput. Appl. Math. **318**, 388–402 (2017)
37. Cornejo, M.E., Medina, J., Ramírez-Poussa, E.: A comparative study of adjoint triples. Fuzzy Sets Syst. **211**, 1–14 (2013)
38. Cornejo, M.E., Medina, J., Ramírez-Poussa, E.: Adjoint negations, more than residuated negations. Inf. Sci. **345**, 355–371 (2016)
39. Davey, B.A., Priestley, H.A.: Introduction to Lattices and Order. Cambridge University Press, Cambridge (2002)
40. Dias, S.M., Vieira, N.J.: Concept lattices reduction: Definition, analysis and classification. In: Expert Syst. Appl., https://doi.org/10.1016/j.eswa.2015.04.044

41. Diáz-Moreno, J.C., Medina, J., Ojeda-Aciego, M.: On basic conditions to generate multi-adjoint concept lattices via Galois connections. Int. J. Gen. Syst. **43**(2), 149–161 (2014)
42. Dubois, D., Prade, H.: Possibility theory and formal concept analysis: Characterizing independent sub-contexts. Fuzzy Sets Syst. **196**, 4–16 (2012)
43. Frič, R., Papčo, M.: A categorical approach to probability theory. Stud. Logica **94**(2), 215–230 (2010)
44. Ganter, B., Kuznetsov, S.: Pattern structures and their projections. In: Delugach, H.S., Stumme, G. (eds.) ICCS 2001. LNCS, vol. 2120, pp. 129–142. Springer, Heidelberg (2001)
45. Ganter, B., Wille, R.: Formal Concept Analysis: Mathematical Foundation. Springer, Heidelberg (1999)
46. Garciá-Pardo, F., Cabrera, I.P., Cordero, P., Ojeda-Aciego, M., Rodríguez-Sanchez, F.J.: On the definition of suitable orderings to generate adjunctions over an unstructured codomain. Inf. Sci. **286**, 173–187 (2014)
47. Georgescu, G., Popescu, A.: Concept lattices and similarity in non-commutative fuzzy logic. Fundamenta Informaticae **53**(1), 23–54 (2002)
48. Halaš, R., Mesiar, R., Pócs, J.: Description of sup- and inf-preserving aggregation functions via families of clusters in data tables, Inf. Sci., to appear. https://doi.org/10.1016/j.ins.2017.02.060
49. Halaš, R., Pócs, J.: Generalized one-sided concept lattices with attribute preferences. Inf. Sci. **303**, 50–60 (2015)
50. Kardoš, F., Pócs, J., Pócsova, J.: On concept reduction based on some graph properties. Knowl.-Based Syst. **93**, 67–74 (2016)
51. Kiseliová, T., Krajči, S.: Generation of representative symptoms based on fuzzy concept lattices. Adv. Soft Computing **33**, 349–354 (2005)
52. Klimushkin, M., Obiedkov, S., Roth, C.: Approaches to the selection of relevant concepts in the case of noisy data. In: Kwuida, L., Sertkaya, B. (eds.) ICFCA 2010. LNCS, vol. 5986, pp. 255–266. Springer, Heidelberg (2010)
53. Konečný, J.: Isotone fuzzy Galois connections with hedges. Inf. Sci. **181**, 1804–1817 (2011)
54. Konečný, J., Medina, J., Ojeda-Aciego, M.: Multi-adjoint concept lattices with heterogeneous conjunctors and hedges. Ann. Math. Artif. Intell. **72**(1–2), 73–89 (2014)
55. Konečný, J., Osička, P.: Triadic concept lattices in the framework of aggregation structures. Inf. Sci. **279**, 512–527 (2014)
56. Krajči, S.: Cluster based efficient generation of fuzzy concepts. Neural Netw. World **13**(5), 521–530 (2003)
57. Krajči, S.: The basic theorem on generalized concept lattice. In Bělohlávek R., Snášel V. (eds.) CLA 2004. CEUR Workshop Proceedings, vol. 110, pp. 25–33. CEUR-WS.org (2004)
58. Krajči, S.: A generalized concept lattice. Log. J. IGPL **13**(5), 543–550 (2005)
59. Krajči, S.: Every concept lattice with hedges is isomorphic to some generalized concept lattice. In Bělohlávek R., Snášel V. (eds.) CLA 2005. CEUR Workshop Proceedings, vol. 162, pp. 1–9. CEUR-WS.org (2005)
60. Krajči, S.: Social Network and Formal Concept Analysis. In: Pedrycz, W., Chen, S.M. (eds.) Social Networks: A framework of Computational Intelligence. Studies in Computational Intelligence, vol 526, pp. 41–62. Springer (2014)
61. Krajči, S., Krajčiová, J.: Social network and one-sided fuzzy concept lattices. In: Spyropoulos, C. (eds.) Fuzz-IEEE 2007. Proceedings of the IEEE International Conference on Fuzzy Systems, pp. 1–6. IEEE Press (2007)
62. Krídlo, O., Krajči, S., Antoni, L.: Formal concept analysis of higher order. Int. J. Gen. Syst. **45**(2), 116–134 (2016)
63. Krídlo, O., Krajči, S., Ojeda-Aciego, M.: The category of L-Chu correspondences and the structure of L-bonds. Fund. Inform. **115**(4), 297–325 (2012)
64. Krídlo, O., Ojeda-Aciego, M.: On L-fuzzy Chu correspondences. Int. J. Comput. Math. **88**(9), 1808–1818 (2011)
65. Krídlo, O., Ojeda-Aciego, M.: Revising the link between L-Chu correspondences and completely lattice L-ordered sets. Ann. Math. Artif. Intell. **72**(1–2), 91–113 (2014)

66. Krupka, M.: On complexity reduction of concept lattices: three counterexamples. Inf. Retr. **15**(2), 151–156 (2012)
67. Kuznetsov, S.O.: On stability of a formal concept. Ann. Math. Artif. Intell. **49**, 101–115 (2007)
68. Kwuida, L., Missaoui, R., Ben Amor, B., Boumedjout, L., Vaillancourt, J.: Restrictions on concept lattices for pattern management. In: Kryszkiewicz, M., Obiedkov, S. (eds.) CLA 2010. CEUR Workshop Proceedings, vol. 672, pp. 235–246. CEUR-WS.org (2010)
69. Medina, J., Ojeda-Aciego, M.: Multi-adjoint t-concept lattices. Inf. Sci. **180**(5), 712–725 (2010)
70. Medina, J., Ojeda-Aciego, M.: On multi-adjoint concept lattices based on heterogeneous conjunctors. Fuzzy Sets Syst. **208**, 95–110 (2012)
71. Medina, J., Ojeda-Aciego, M.: Dual multi-adjoint concept lattices. Inf. Sci. **225**, 47–54 (2013)
72. Medina, J., Ojeda-Aciego, M., Pócs, J., Ramiréz-Poussa, E.: On the Dedekind-MacNeille completion and formal concept analysis based on multilattices. Fuzzy Sets Syst. **303**, 1–20 (2016)
73. Medina, J., Ojeda-Aciego, M., Ruiz-Calviño, J.: On multi-adjoint concept lattices: definition and representation theorem. Lect. Notes Artif. Intell **4390**, 197–209 (2007)
74. Medina, J., Ojeda-Aciego, M., Ruiz-Calviño, J.: Relating generalized concept lattices and concept lattices for non-commutative conjunctors. Appl. Math. Lett. **21**, 1296–1300 (2008)
75. Medina, J., Ojeda-Aciego, M., Ruiz-Calviño, J.: Formal concept analysis via multi-adjoint concept lattices. Fuzzy Sets Syst. **160**(2), 130–144 (2009)
76. Medina, J., Ojeda-Aciego, M., Valverde, A., Vojtáš, P.: Towards biresiduated multi-adjoint logic programming. In: Conejo, R., Urretavizcaya, M., Pérez-de-la-Cruz, J.-L. (eds.) CAEPIA-TTIA 2003. LNCS (LNAI), vol. 3040, pp. 608–617. Springer, Heidelberg (2004)
77. Medina, J., Ojeda-Aciego, M., Vojtáš, P.: Similarity-based unification: a multi-adjoint approach. Fuzzy Sets Syst. **146**, 43–62 (2004)
78. Ore, Ø.: Galois connexions. Trans. Am. Mathe. Soc. **55**, 493–513 (1944)
79. Pócs, J.: Note on generating fuzzy concept lattices via Galois connections. Inf. Sci. **185**(1), 128–136 (2012)
80. Pócs, J.: On possible generalization of fuzzy concept lattices using dually isomorphic retracts. Inf. Sci. **210**, 89–98 (2012)
81. Pócs, J., Pócsová, J.: On some general aspects of forming fuzzy concept lattices. Appl. Math. Sci. **7**(112), 5559–5605 (2013)
82. Poelmans, J., Dedene, G., Verheyden, G., Van der Mussele, H., Viaene, S., Peters, E.: Combining business process and data discovery techniques for analyzing and improving integrated care pathways. In: Perner, P. (ed.) ICDM 2010, pp. 505–517. Springer-Verlag, Berlin Heidelberg (2010)
83. Poelmans, J., Ignatov, D.I., Kuznetsov, S.O., Dedene, G.: Formal concept analysis in knowledge processing: a survey on applications. Expert Syst. Appl. **40**(16), 6538–6560 (2013)
84. Poelmans, J., Kuznetsov, S.O., Ignatov, D.I., Dedene, G.: Formal concept analysis in knowledge processing: a survey on models and techniques. Expert Syst. Appl. **40**, 6601–6623 (2013)
85. Pollandt, S.: Fuzzy Begriffe. Springer, Berlin (1997)
86. Popescu, A.: A general approach to fuzzy concepts. Math. Logic Quart. **50**(3), 265–280 (2004)
87. Priss, U.: Using FCA to Analyse How Students Learn to Program. In: Cellier, P., Distel, F., Ganter, B. (eds.) ICFCA 2013. LNCS, vol. 7880, pp. 216–227. Springer, Heidelberg (2013)
88. Singh, P.K., Kumar, ChA: Bipolar fuzzy graph representation of concept lattice. Inf. Sci. **288**, 437–448 (2014)
89. Singh, P.K., Kumar, ChA: A note on bipolar fuzzy graph representation of concept lattice. Int. J. Comp. Sci. Math. **5**(4), 381–393 (2014)
90. Skřivánek, V., Frič, R.: Generalized random events. Int. J. Theor. Phys. **54**, 4386–4396 (2015)
91. Snášel, V., Duráková, D., Krajči, S., Vojtáš, P.: Merging concept lattices of α-cuts of fuzzy contexts. Contrib. Gen. Algebra **14**, 155–166 (2004)
92. Valverde-Albacete, F.J., Peláez-Moreno, C.: Towards a generalisation of formal concept analysis for data mining purposes. In: Missaoui, R., Schmid, J. (eds.) ICFCA 2006. LNCS, vol. 3874, pp. 161–176. Springer, Heidelberg (2006)

93. Valverde-Albacete, F.J., Peláez-Moreno, C.: Extending conceptualisation modes for generalised formal concept analysis. Inf. Sci. **181**, 1888–1909 (2011)
94. Wille, R.: Restructuring lattice theory: an approach based on hierarchies of concepts. In: Rival, I. (ed.) Ordered Sets, pp. 445–470. Reidel, Dodrecht-Boston (1982)

Generating Fuzzy Attribute Rules Via Fuzzy Formal Concept Analysis

Valentín Liñeiro-Barea, Jesús Medina and Inmaculada Medina-Bulo

Abstract Extracting knowledge from databases is a procedure which interest has increased in a wide variety of areas like stock market, medicine or census data, to name a few. A compact representation of this knowledge is given by rules. Formal concept analysis plays an important role in this area. This paper introduces a new kind of attribute implications considering the fuzzy notions of support and confidence and is also focused on the particular case in which the set of attributes are intensions. Moreover, an application to clustering for size reduction of concept lattices is included.

Keywords Fuzzy rule · Fuzzy formal concept analysis · Fuzzy set

1 Introduction

The hidden knowledge databases store is a valuable asset in a wide variety of areas like stock market prediction [16, 22], disease diagnosis [20, 21] or census data analysis [7, 18], among others. One of the main techniques in order to represent the knowledge extracted from a database is by rules, summarizing completely the information stored in the database. These rules are usually extracted via APRIORI

Partially supported by the State Research Agency (AEI) and the European Regional Development Fund (ERDF) projects TIN2015-65845-C3-3-R and TIN2016-76653-P.

V. Liñeiro-Barea · J. Medina (✉)
Department of Mathematics, University of Cádiz, Cadiz, Spain
e-mail: valentin.lineiro@uca.es

J. Medina
e-mail: jesus.medina@uca.es

I. Medina-Bulo
Department of Computer Engineering, University of Cádiz, Cadiz, Spain
e-mail: inmaculada.medina@uca.es

© Springer International Publishing AG 2018
L. T. Kóczy and J. Medina (eds.), *Interactions Between Computational Intelligence and Mathematics*, Studies in Computational Intelligence 758,
https://doi.org/10.1007/978-3-319-74681-4_7

algorithms [1], which explore frequent itemsets to select the most frequent and confident rules of the database.

Formal Concept Analysis (FCA) [8] is a mathematical technique which helps to discover relationships between sets of attributes and objects inside a database, known as concepts. FCA retrieves the main concepts that a database has, and that can be useful to obtain a rule set. In the classical FCA [1], only boolean attributes are considered, leading to a set of rules which consider attributes fully true or false. A most accurate framework is given by fuzzy generalizations where uncertainty and noise are present.

The computation of attribute implications in one of the most important research topics in FCA [4, 5, 10, 11, 15, 28]. The main goal of this paper is to define, in any fuzzy concept lattice framework, a new kind of fuzzy attribute implications (that is a rule with attributes that might not be fully true or false) and a base of them, as a minimum set of rules needed to summarize the whole information present in a database. A particular case in which intensions are considered in the rules is studied and an application to clustering the concepts of the concept lattice, which provides a size reduction mechanism of the concept lattice, is presented.

This paper is structured as follows. In Sect. 2 the preliminary definitions are presented. Then, Sect. 3 presents fuzzy sc-attribute rules, properties and the notion of base. Next, Sect. 4 studies the particular case in which intensions are considered in the rules and Sect. 5 applies these rules in order to provide a clustering in the concept lattice. Finally, conclusions and future work are shown.

2 Preliminaries

This section presents the required preliminary definitions. In all the definitions a complete lattice $(L, \preceq)$ and a finite universe U are considered.

The following definition introduces the notion of spanning tree, which will be used in Sect. 4.

Definition 1 ([24]) A *spanning tree* S of a connected graph $\langle V, E \rangle$ is a minimal subset of E which interconnects all elements in V.

The notion of cardinality is given in a fuzzy framework as follows.

Definition 2 ([31]) Given a fuzzy set $f : U \to L$, the fuzzy cardinality of f defined as $\text{card}(f) = \sum_{i \in U} f(i)$ is called *sigma count*.

Formal Concept Analysis (FCA) is applied to relational databases in order to extract pieces of information, which are called concepts, and they are hierarchized forming a complete lattice. The concepts are obtained from two concept-forming operators, which form a Galois connection. There exist different extensions of formal concept analysis to a fuzzy framework and almost all of them are based on a context (a set of attributes A, a set of objects B and a fuzzy relation R between them) and a Galois connection [3, 14, 19] or a family of Galois connections [2, 23].

The theory and results proposed in this paper are applied to all of them. In order to recall a particular fuzzy concept lattice, mainly for the examples, we will present the notion of context and the definitions of the concept-forming operators introduced in this framework.

The first definition introduces the basic operators considered in the concept-forming operators.

Definition 3 Let $(P_1, \leq_1)$, $(P_2, \leq_2)$, $(P_3, \leq_3)$ be posets and $\&: P_1 \times P_2 \to P_3$, $\swarrow: P_3 \times P_2 \to P_1$, $\nwarrow: P_3 \times P_1 \to P_2$ be mappings, then$(\&, \swarrow, \nwarrow)$ is an *adjoint triple* with respect to P_1, P_2, P_3 if:

$$\text{Adjoint property:} \quad x \leq_1 z \swarrow y \quad \text{iff} \quad x \& y \leq_3 z \quad \text{iff} \quad y \leq_2 z \nwarrow x$$

where $x \in P_1$, $y \in P_2$ and $z \in P_3$.

The posets $(P_1, \leq_1)$ and $(P_2, \leq_2)$ considered in the previous definition are complete lattices in the multi-adjoint concept lattice framework [19] and they are denoted as $(L_1, \leq_1)$ and $(L_2, \leq_2)$. Now the notion of multi-adjoint frame is presented.

Definition 4 A multi-adjoint frame $\mathcal{L}$ is a tuple

$$(L_1, L_2, P, \leq_1, \leq_2, \leq, \&_1, \swarrow^1, \nwarrow_1, \ldots, \&_n, \swarrow^n, \nwarrow_n)$$

where $(L_1, \leq_1)$ and $(L_2, \leq_2)$ are complete lattices, $(P, \leq)$ is a poset and $(\&_i, \swarrow^i, \nwarrow_i)$ is an adjoint triple with respect to L_1, L_2, P, for all $i \in \{1, \ldots, n\}$.

Given a multi-adjoint frame, a context can be introduced.

Definition 5 Let $(L_1, L_2, P, \&_1, \ldots, \&_n)$ be a multi-adjoint frame, a context is a tuple (A, B, R, σ) such that A and B are non-empty sets (interpreted as attributes and objects, respectively), R is a P-fuzzy relation $R: A \times B \to P$ and $\sigma: A \times B \to \{1, \ldots, n\}$ is a mapping which associates any element in $A \times B$ with a particular adjoint triple in the frame.

The symbols L_2^B and L_1^A denote the set of fuzzy subsets $g: B \to L_2$, $f: A \to L_1$ respectively. On these sets a pointwise partial order can be considered from the partial orders in $(L_1, \leq_1)$ and $(L_2, \leq_2)$, which provides L_2^B and L_1^A the structure of complete lattice.

After introducing the notions of multi-adjoint frame and context are introduced, the definition of the concept-forming operators in this framework can be given. The mappings $\uparrow: L_2^B \to L_1^A$ and $\downarrow: L_1^A \to L_2^B$ are defined as:

$$g^{\uparrow}(a) = \inf\{R(a, b) \swarrow^{\sigma(a,b)} g(b) \mid b \in B\}$$
$$f^{\downarrow}(b) = \inf\{R(a, b) \nwarrow_{\sigma(a,b)} f(a) \mid a \in A\}$$

Table 1 Relation R in Example 1

R	a_1	a_2	a_3
b_1	0.5	0.0	1.0
b_2	1.0	0.5	0.0

for all $g \in L_2^B$, $f \in L_1^A$, $a \in A$, $b \in B$. These two operators form a Galois connection [19]. The notion of concept can be defined as usual: A multi-adjoint concept is a pair $\langle g, f \rangle$ satisfying that $g \in L_2^B$, $f \in L_1^A$ and $g^\uparrow = f$, $f^\downarrow = g$, with $(\uparrow, \downarrow)$ being the Galois connection defined above. The fuzzy subsets g and f in a concept are usually known as the extent and intent of the concept, respectively.

Definition 6 The multi-adjoint concept lattice associated with a multi-adjoint frame $(L_1, L_2, P, \&_1, \ldots, \&_n)$ and a context (A, B, R, σ) is the set

$$\mathcal{M} = \{\langle g, f \rangle \mid g \in L_2^B, f \in L_1^A \text{ and } g^\uparrow = f, f^\downarrow = g\}$$

in which the ordering is defined by $\langle g_1, f_1 \rangle \preceq \langle g_2, f_2 \rangle$ if and only if $g_1 \preceq g_2$ ($f_2 \preceq f_1$). This ordering provides $\mathcal{M}$ the structure of a complete lattice [19].

Example 1 Let $(L, \preceq, \&_G)$ be a multi-adjoint frame, where $\&_G$ is the Gödel conjunctor with respect to $L = \{0, 0.5, 1\}$. The context (A, B, R, σ) is formed by the sets $A = \{a_1, a_2, a_3\}$ and $B = \{b_1, b_2\}$, the relation R is defined from Table 1 and the constant mapping σ. In the fuzzy set notation, for each $a \in A$, the expression $a/1$ will simply be written as a and $a/0$ will be omitted.

The concept lattice related to the given context is shown in Fig. 1. We can see we have seven concepts, and the Hasse diagram shows the hierarchy among them.

3   Fuzzy sc-Attribute Rules

This section presents a special kind of fuzzy attribute rule and several properties. From now on, a frame and a context (A, B, R, σ) will be fixed from which a Galois connection $(\uparrow, \downarrow)$ or a family of Galois connections (as in [2, 6, 23]) is defined. Specifically, we will consider a Galois connection instead of a family, in order to simplify the notation.

First of all, the classical definitions of support and confidence are provided.

Definition 7 ([1]) The support is defined as the ratio of objects which are related to a given subset of attributes.

Definition 8 ([1]) Given a set of attributes A and subsets $Y_1, Y_2 \in A$, the confidence is defined as the probability of, given any object that has the attributes in Y_1, it also has the attributes of Y_2.

Fig. 1 Concept lattice
related to the context given
in Example 1

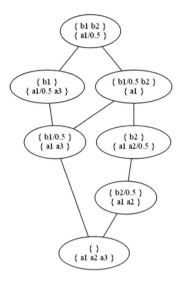

These measures on rules will be extended to the fuzzy one next. The first notion
we will consider is the definition of support, which is a key measure of a fuzzy rule.

Definition 9 The support of $f \in L_1^A$ in (A, B, R, σ) is defined as

$$\text{supp}(f) = \frac{\text{card}(f^{\downarrow})}{|B|} \qquad (1)$$

This definition is applied in the following example.

Example 2 Continuing with Example 1, the support of the intent $\{a_1/0.5, a_3\}$ is
given by:

$$\text{supp}(\{a_1/0.5, a_3\}) = \frac{\text{card}(\{a_1/0.5, a_3\}^{\downarrow})}{|B|}$$
$$= \frac{\text{card}(b_1)}{2} = \frac{1}{2}$$

The support satisfies interesting properties that are shown next. Since the proof
can straightforwardly be obtained, it is not included.

Proposition 1 *Given $f_1, f_2, f_{\perp} \in L_1^A$, with $f_{\perp}(a) = \perp$ for all $a \in A$, the following
properties hold:*

1. *If $f_1 \leq f_2$ then $\text{supp}(f_2) \leq \text{supp}(f_1)$.*
2. *$\text{supp}(f_{\perp}) = 1$.*
3. *$0 \leq \text{supp}(f_1) \leq 1$.*

Once the notion of support has been introduced, the following definition provides the definition of fuzzy sc-attribute rule and a specific truth value associated with it.

Definition 10 Given two fuzzy subsets of attributes f_1, $f_2 \in L_1^A$, the *fuzzy attribute rule over A from f_1 to f_2* is given by the expression $f_2 \leftarrow_{(s,c)} f_1$, where $s = \text{supp}(f_1)$ and the confidence c is defined by

$$c = \frac{\text{supp}(f_1 \cup f_2)}{\text{supp}(f_1)} \tag{2}$$

If the confidence is 1, the fuzzy sc-attribute rule is called *fuzzy sc-attribute implication*, which is denoted by $f_2 \Leftarrow f_1$.

Now, an example of the application of the notion of confidence will be shown.

Example 3 From Example 1, the fuzzy attribute rule $\{a_1\} \leftarrow_{(s,c)} \{a_1/0.5\}$ can be considered, where the confidence is computed as:

$$c = \frac{\text{supp}(\{a_1/0.5\} \cup \{a_1\})}{\text{supp}(\{a_1/0.5\})}$$
$$= \frac{\text{supp}(\{a_1\})}{\text{supp}(\{a_1/0.5\})} = \frac{\frac{\text{card}(\{a_1\}^\downarrow)}{|B|}}{\frac{\text{card}(\{a_1/0.5\}^\downarrow)}{|B|}} = \frac{\text{card}(\{a_1\}^\downarrow)}{\text{card}(\{a_1/0.5\}^\downarrow)}$$
$$= \frac{\text{card}(\{b_1/0.5, b_2\})}{\text{card}(\{b_1, b_2\})} = \frac{1.5}{2} = 0.75$$

This mapping satisfies several interesting properties as the following one, which trivially holds.

Proposition 2 *The confidence c of a fuzzy attribute rule verify $0 \le c \le 1$.*

The following section will present different properties of the introduced rules.

3.1 Properties of the Fuzzy sc-Attribute Rules

Given two ordering related fuzzy subsets of attributes, a trivial fuzzy attribute rule always arises.

Proposition 3 *Let f_1, $f_2 \in L_1^A$, where $f_1 \prec f_2$, and the rule $f_1 \leftarrow_{(s,c)} f_2$, we have that $c = 1$.*

Proof The confidence of the rule can be expressed as $\frac{\text{supp}(f_1 \cup f_2)}{\text{supp}(f_2)} = \frac{\text{supp}(f_2)}{\text{supp}(f_2)} = 1$. $\square$

A relation in the subsets of attributes implies a relation in the confidences.

Proposition 4 Let $f_1, f_2, f_3 \in L_1^A$, where $f_1 \preceq f_2$ and $f_1 \preceq f_3$, and the rules $f_2 \leftarrow_{(s_{12}, c_{12})} f_1, f_3 \leftarrow_{(s_{13}, c_{13})} f_1$. If $f_2 \preceq f_3$, then $c_{13} \leq c_{12}$.

Proof The confidences c_{13} and c_{12} arise from Eq. 2:

$$c_{12} = \frac{\text{supp}(f_1 \cup f_2)}{\text{supp}(f_1)} = \frac{\text{supp}(f_2)}{\text{supp}(f_1)}$$
$$c_{13} = \frac{\text{supp}(f_1 \cup f_3)}{\text{supp}(f_1)} = \frac{\text{supp}(f_3)}{\text{supp}(f_1)}$$

Assuming that $f_2 \preceq f_3$, we also know that $\text{supp}(f_3) \leq \text{supp}(f_2)$. Applying this fact, the following inequalities are equivalents:

$$\text{supp}(f_3) \leq \text{supp}(f_2)$$
$$\frac{\text{supp}(f_3)}{\text{supp}(f_1)} \leq \frac{\text{supp}(f_2)}{\text{supp}(f_1)}$$
$$c_{13} \leq c_{12}$$

This establishes that, if $f_2 \preceq f_3$, then $c_{13} \leq c_{12}$. ☐

Example 4 In the context of Example 1, we will see the previous property in the rules $\{a_1\} \leftarrow_{(s_1, c_1)} \{a_1/0.5\}$ and $\{a_1, a_3\} \leftarrow_{(s_2, c_2)} \{a_1/0.5\}$. Hence, we compute the supports and confidences of both rules:

$$s_1 = \text{supp}(\{a_1/0.5\}) = 1$$
$$c_1 = \frac{\text{supp}(\{a_1\})}{\text{supp}(\{a_1/0.5\})} = \frac{0.75}{1} = 0.75$$
$$s_2 = \text{supp}(\{a_1/0.5\}) = 1$$
$$c_2 = \frac{\text{supp}(\{a_1, a_3\})}{\text{supp}(\{a_1/0.5\})} = \frac{0.25}{1} = 0.25$$

As we can see, $c_2 \leq c_1$. Now, we will consider other two rules that do not satisfy the hypotheses in Proposition 4 and we will see that the thesis does not hold either. Given the rules $\{a_1, a_3\} \leftarrow_{(s_3, c_3)} \{a_1\}$ and $\{a_1, a_2/0.5\} \leftarrow_{(s_4, c_4)} \{a_1\}$, we have:

$$s_3 = \text{supp}(\{a_1\}) = 0.75$$
$$c_3 = \frac{\text{supp}(\{a_1, a_3\})}{\text{supp}(\{a_1\})} = \frac{0.25}{0.75} = 0.33$$
$$s_4 = \text{supp}(\{a_1\}) = 0.75$$
$$c_4 = \frac{\text{supp}(\{a_1, a_2/0.5\})}{\text{supp}(\{a_1/0.5\})} = \frac{0.5}{0.75} = 0.67$$

which implies that $c_4 \not\preceq c_3$.

From Proposition 4, the following corollary arises.

Corollary 1 *Let $f_1, f_2, f_3 \in L_1^A$, where $f_1 \prec f_2 \prec f_3$, and the rules $f_2 \leftarrow_{(s,c)} f_1$, $f_3 \leftarrow_{(s',1)} f_1$, we have that $c = 1$.*

Proof Since $f_2 \prec f_3$, by Proposition 4, we obtain that $1 \leq c$, therefore, as 1 is the greatest possible confidence, $c = 1$ holds.   □

The following result shows how the confidence can be derived via transitivity, based on the idea given in [17].

Theorem 1 *Let $f_1, f_2, f_3 \in L_1^A$, where $f_1 \prec f_2 \prec f_3$, and the rules $f_2 \leftarrow_{(s,c)} f_1$, $f_3 \leftarrow_{(s',c')} f_2$, $f_3 \leftarrow_{(s,c'')} f_1$, then we have that $c \cdot c' = c''$.*

Proof Let $f_3 \leftarrow_{(s'',c'')} f_1$. The support of the new rule, s'' can be obtained applying Eq. 1:

$$s'' = \text{supp}(f_1) = s$$

The confidence of the new rule c'' arises from Eq. 2:

$$c'' = \frac{\text{supp}(f_1 \cup f_3)}{\text{supp}(f_1)} = \frac{\text{supp}(f_3)}{\text{supp}(f_1)}$$

Applying again Eq. 2, c and c' are actually:

$$c = \frac{\text{supp}(f_1 \cup f_2)}{\text{supp}(f_1)} = \frac{\text{supp}(f_2)}{\text{supp}(f_1)}$$

$$c' = \frac{\text{supp}(f_2 \cup f_3)}{\text{supp}(f_2)} = \frac{\text{supp}(f_3)}{\text{supp}(f_2)}$$

If c and c' are multiplied:

$$c \cdot c' = \frac{\text{supp}(f_2)}{\text{supp}(f_1)} \cdot \frac{\text{supp}(f_3)}{\text{supp}(f_2)} = \frac{\text{supp}(f_3)}{\text{supp}(f_1)} = c''$$

This establishes that $s'' = s$ and $c'' = c \cdot c'$.   □

Finally, the following theorem extends the previous transitivity property.

Theorem 2 *Let $f_1, f_2, f_3, f_4 \in L_1^A$, where $f_1 \preceq f_2$, $f_1 \preceq f_3$ and $f_3 \preceq f_4$, $f_2 \preceq f_4$ and the rules $f_2 \leftarrow_{(s_1,c_1)} f_1$, $f_3 \leftarrow_{(s_2,c_2)} f_1$, $f_4 \leftarrow_{(s_3,c_3)} f_3$ and $f_4 \leftarrow_{(s_4,c_4)} f_2$, then we obtain that $c_4 = c_2 \cdot c_3 \cdot \frac{1}{c_1}$.*

Proof Let $f_4 \leftarrow_{(s_4,c_4)} f_2$. The confidence of the new rule c_4 arises from Eq. 2:

$$c_4 = \frac{\text{supp}(f_2 \cup f_4)}{\text{supp}(f_2)} = \frac{\text{supp}(f_4)}{\text{supp}(f_2)}$$

In other hand, we know that:

$$
\begin{aligned}
c_2 \cdot c_3 \cdot \frac{1}{c_1} &= \frac{\text{supp}(f_1 \cup f_3)}{\text{supp}(f_1)} \cdot \frac{\text{supp}(f_3 \cup f_4)}{\text{supp}(f_3)} \cdot \frac{\text{supp}(f_1)}{\text{supp}(f_1 \cup f_2)} \\
&= \frac{\text{supp}(f_3)}{\text{supp}(f_1)} \cdot \frac{\text{supp}(f_4)}{\text{supp}(f_3)} \cdot \frac{\text{supp}(f_1)}{\text{supp}(f_2)} = \frac{\text{supp}(f_4)}{\text{supp}(f_2)} = c_4
\end{aligned}
$$

This establishes that $c_4 = c_2 \cdot c_3 \cdot \frac{1}{c_1}$. $\qquad\square$

This property allows to remove cycles in the concept lattice, which will be fundamental in order to compute a minimal base of fuzzy sc-attribute rules. Note that a concept lattice can be considered as a graph, in which the concepts are the vertices and the edges are given by the relations among them in the Hasse diagram of the concept lattice. This relation will be considered throughout this paper.

3.2 Fuzzy Sc-Attribute Rule Base

Given a context, the set of fuzzy attribute rules computed may be huge and it also may include redundant and not interesting rules for the purpose they are being obtained. In order to fix that, a base of rules should be obtained.

Definition 11 A fuzzy attribute rule $f' \leftarrow_{(s,c)} f$ is *derived* from a set of fuzzy attribute rules M if a succession $\{f_1, f_2, \cdots, f_n\}$ of different fuzzy subset of attributes exists, such that $f = f_1$, $f' = f_n$, and $f_{i+1} \leftarrow_{(s_i,c_i)} f_i \in M$ or $f_i \leftarrow_{(s_i,c_i)} f_{i+1} \in M$, for all $i \in \{1, \ldots, n-1\}$.

Example 5 Considering in Example 1 the set of fuzzy attribute rules $M = \{\{a_1\} \leftarrow_{(s_{12}=1,c_{12}=0.75)} \{a_1/0.5\}, \{a_1, a_3\} \leftarrow_{(s_{23}=0.75,c_{23}=0.\hat{3})} \{a_1\}\}$, we can derive a new rule from it, $\{a_1, a_3\} \leftarrow_{(s,c)} \{a_1/0.5\}$, with:

$$s = s_{12} = 1$$
$$c = c_{12} \cdot c_{23} = 0.75 \cdot 0.\hat{3} = 0.25$$

From the notion of derivation, the definition of minimal base is introduced.

Definition 12 A set of fuzzy rules T is a *minimal base* if the following properties hold:

- Completeness. Any rule $f_2 \leftarrow_{(s,c)} f_1 \notin T$ can be derived from the rules in T.
- Non-redundancy. No rule $f_2 \leftarrow_{(s,c)} f_1 \in T$ can be derived from the rest of rules in T.

A minimal base of fuzzy rules T is formed by two special subsets, depending on the confidence threshold considered: the set of fuzzy rules with confidence $c = 1$, which is called *fuzzy attribute implication base* and, the set of rules with $c < 1$, which is called *fuzzy sc-attribute rule base*.

Current papers in fuzzy attribute implications [26, 27] only consider fuzzy rules that have the greatest confidence value ($c = 1$), however considering implication with $c < 1$ is also interesting since some noise in the data can be mitigated and outlier data are also considered. The following section will be focused on this kind of implications.

4   Fuzzy sc-Intension Rules

This section defines, in a given concept lattice framework, an sc-attribute rule base extending the classical case [17] and relates this base to the given concept lattice. First of all, the notion of fuzzy sc-intension rule is introduced.

Definition 13 Given the set of all intents in the context $\mathrm{Int}(A, B, R, \sigma) = \{f \in L_1^A \mid f = f^{\downarrow\uparrow}\}$, the fuzzy rule $f_2 \leftarrow_{(s,c)} f_1$, in which the fuzzy subset of attributes are intents, that is $f_1, f_2 \in \mathrm{Int}(A, B, R, \sigma)$, and $f_1 \prec f_2$ is called *fuzzy sc-intension rule*.

Note that Theorems 1 and 2 show how new fuzzy sc-intension rules can be derived from a base via transitivity. For example, given $f_1, f_2, f_3 \in \mathrm{Int}(A, B, R, \sigma)$, satisfying $f_1 \prec f_2 \prec f_3$, and the rules $f_2 \leftarrow_{(s,c)} f_1$ and $f_3 \leftarrow_{(s',c')} f_2$, the rule $f_3 \leftarrow_{(s,c'')} f_1$ can straightforwardly be derived considering $c'' = c \cdot c'$.

This result provides that only the rules between neighbor intents need to be considered in a base and gives us a mechanism in order to compute a fuzzy sc-intension rule base, only considering a minimal subset of rules between neighbor intents removing cycles in the concept lattice, in other words, obtaining a spanning tree of the Hasse diagram of the concept lattice.

If we need to compute the confidence of a new rule, we just need to identify the path followed in the spanning tree and then multiply the confidences of the involved rules, inverting the value in the rules that we consider in the reverse form. This is shown in the following example.

Example 6 Given a spanning tree $\mathcal{S}$ of a concept lattice $\mathcal{M}$ formed by the partial implications $\{f_2 \leftarrow_{(0.5,0.4)} f_1, f_3 \leftarrow_{(0.4,0.3)} f_1, f_4 \leftarrow_{(0.2,0.25)} f_2\}$, the derived partial implication $f_4 \leftarrow_{(s,c)} f_3$ can be obtained transiting in the spanning tree as follows: First of all, we go from the node related to f_3 to f_1 considering the rule $f_3 \leftarrow_{(0.4,0.3)}$

f_1, then we go from f_1 to f_2 by $f_2 \leftarrow_{(0.5,0.4)} f_1$ and, finally, we pass from f_2 to f_3 considering the rule $f_4 \leftarrow_{(0.2,0.25)} f_2$.

Hence, the support of the derived rule is trivially supp(f_3) and the confidence is $\frac{1}{0.3} \cdot 0.4 \cdot 0.25 = 0.\hat{3}$.

Once a mechanism for deriving rules has been presented, the definition of fuzzy sc-intension rule based is introduced.

Definition 14 A subset $T_L \subseteq \{f_2 \leftarrow_{(s,c)} f_1 \mid f_1, f_2 \in \text{Int}(A, B, R, \sigma)$ and $f_1 \prec f_2\}$ is called *fuzzy sc-intension rule base of the context* if it is a minimal base.

As it was commented previously, this base is related to the notion of spanning tree.

Proposition 5 *Given a fuzzy sc-intension rule base of the context T_L, the graph $\langle V, E \rangle$, where $V = \text{Int}(A, B, R, \sigma)$ and $E = \{(f_1, f_2) \mid f_2 \leftarrow_{(s,c)} f_1 \in T_L\}$, is a spanning tree of the concept lattice $\mathcal{M}$ associated with the context.*

Since multiple spanning trees of a concept lattice $\mathcal{M}$ may exist, T_L is not unique, although the size of the base is fixed, as the following result shows.

Theorem 3 *Given the concept lattice $\mathcal{M}$ associated with the context (A, B, R, σ), and a fuzzy sc-intension rule base T_L of the context, we have that its size is always $|\mathcal{M}| - 1$.*

Proof Applying Definition 14, the set of edges of the rules in the base forms a spanning tree of $\mathcal{M}$. As the number of edges of any spanning tree of a graph is $|V| - 1$, begin V the number of nodes in the graph, the size of the spanning tree associated with $\mathcal{M}$ is $|\mathcal{M}| - 1$. □

The most interesting spanning trees are those associated with the rules with bigger confidence. Given a minimal base T_L, the spanning tree related to T_L, in which each edge is labeled with the confidence of the corresponding fuzzy sc-intension rule, will be called T_L-*weighted spanning tree*.

5 Application to Clustering

The base of fuzzy sc-intension rules has immediate application in a clustering concept task. Given a concept lattice $\mathcal{M}$ associated with a context (A, B, R), we can naturally see the confidence of a rule $f_2 \leftarrow_{(s,c)} f_1$ as a ratio of similarity between the concepts $C_2 = \langle f_2^{\downarrow}, f_2 \rangle$ and $C_1 = \langle f_1^{\downarrow}, f_1 \rangle$. If we establish a threshold δ of confidence, we can clustered the concepts of $\mathcal{M}$ obtaining a set of the similar ones. Note that one concept can be part of more than one cluster. Hence, the obtained clustering can be a covering [9, 30].

Given a concept C, the set of concepts which has a confidence greater or equal than δ with respect to C will be called as *the cover in $\mathcal{M}$ of C* and will be denoted as $[C]$.

The process of clustering can be summarized in the following steps:

1. Obtaining a fuzzy sc-intension rule base of the context.
2. Traversing the fuzzy rules base from the top as follows:

   – Considering two sets, the set of concepts $\mathcal{M}$ and the set in which the computed clusters will be added, $\mathcal{N}$, which is empty at the first.
   – Step 1. Fix the top concept $C_1 \in \mathcal{M}$.
   – Step 2. Consider in the cover of C_1, $[C_1]$, with respect to δ.
   – Step 3. Add $[C_1]$ to $\mathcal{N}$ and remove the set $[C_1]$ from $\mathcal{M}$.
   – Step 4. After removing $[C_1]$ from $\mathcal{M}$ the rest of concepts form a meet-semilattice. Apply Steps 3–4 to each maximal element $C_{M_1}, \ldots, C_{M_n}$.
   – Step 5. Apply Steps 2–4 until $\mathcal{M}$ be empty.

It is clear that depending on Step 4, the clustering provides a partition or a covering.

(a) If Step 4 is doing in a sequential process, that is, firstly Steps 2 and 3 are applied to the maximal element C_{M_1}, then to C_{M_2}, and so on, a partition of $\mathcal{M}$ is obtained. The main weakness is that this partition depend on the considered ordering in the maximal elements.
(b) If Step 4 is doing in a parallel process, that is, firstly Step 2 is applied to each maximal element $C_{M_1}, \ldots, C_{M_n}$, and the different, not necessary disjoint classes $[C_{M_1}], \ldots, [C_{M_n}]$ are computed, and then Step 3 is applied, a covering is obtained. In this case, a partition is not obtained, in general.

In the following example, the first option of the clustering algorithm is applied to a particular concept lattice.

Example 7 Following Example 1, we obtain the fuzzy sc-intension rule base T_L shown in Fig. 2, which provides a maximal T_L-weighted spanning tree. The confidence value of each rule is indicated in the edges of the diagram. In each node, together with the extension and intension of the corresponding concept, a number is written representing the label of the concept.

If we assume $\delta = 0.66$, we obtain the following partition:

- $[C_1] = \{1, 2\}$
- $[C_3] = \{3\}$
- $[C_4] = \{4\}$
- $[C_5] = \{5\}$
- $[C_6] = \{6\}$
- $[C_7] = \{7\}$

In case of $\delta = 0.5$, we obtain the following partition:

- $[C_1] = \{1, 2, 3, 4\}$
- $[C_5] = \{5\}$
- $[C_6] = \{6\}$
- $[C_7] = \{7\}$

Considering now $\delta = 0.25$, the partition is:

Fig. 2 Fuzzy sc-intension
rule base related to the
context given in Example 1

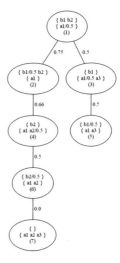

- $[C_1] = \{1, 2, 3, 4, 5, 6\}$
- $[C_7] = \{7\}$

Note that, in this particular case, since a small concept lattice has been considered, the algorithm with the second option (b) provides the same (partition) covering.

An interesting consequence of the proposed clustering mechanism is that it provides a reduction procedure based on the confidence of the fuzzy sc-intension rules. This is possible because the set of classes always contains the minimum concept. Moreover, we have that the concepts closer to the top concept provides greater confidences. Therefore, the computation of the top part of the concept lattice is interesting in order to obtain the most representative fuzzy sc-intension rules.

6 Conclusions and Future Work

A new kind of attribute implication has been introduced considering the fuzzy notions of support and confidence, and the properties of these rules have been studied. These rules allow to consider fuzzy rules that have a confidence value less than one, $c < 1$, which is also interesting since some noise in the data can be mitigated and outlier data are also considered. The particular case of using intensions in the rules provides attribute implications with confidence less than 1, which can complement the usual attribute implications with truth degree 1. Moreover, these rules provide a procedure for clustering the concept lattices and, as a consequence, offer a size reduction mechanism of the concept lattice.

In the following, a set of future work lines will be presented. First of all, the performance and the potential of the procedure for information retrieval will be done with an experimental survey. This will be the main line for future work, allowing to search for optimizations in big data sets.

Extracting information from a database is the main task fuzzy rules are designed for. Fuzzy rules can be utilized in Machine Learning scenarios such classification [13, 29] or prediction [12, 25]. Another line for future work will be to consider such scenarios in order to study the potential of our work in them.

References

1. Agrawal, R., Imielinski, T., Swami, A.: Mining association rules between sets of items in large databases. In: ACM SIGMOD International Conference on Management of Data, pp. 207–216 (1993)
2. Antoni, L., Krajci, S., Kridlo, O., Macek, B., Pisková, L.: On heterogeneous formal contexts. Fuzzy Sets Syst. **234**, 22–33 (2014)
3. Bělohlávek, R.: Lattices of fixed points of fuzzy Galois connections. Math. Logic Q. **47**(1), 111–116 (2001)
4. Belohlavek, R., Cordero, P., Enciso, M., Mora, A., Vychodil, V.: Automated prover for attribute dependencies in data with grades. Int. J. Approx. Reasoning **70**, 51–67 (2016)
5. Belohlávek, R., Vychodil, V.: Attribute dependencies for data with grades ii. Int. J. Gen. Syst. **46**(1), 66–92 (2017)
6. Butka, P., Pócs, J.: Generalization of one-sided concept lattices. Comput. Inform. **32**(2), 355–370 (2013)
7. Chertov, O., Aleksandrova, M.: Using association rules for searching levers of influence in census data. Procedia Soc. Behav. Sci. **73**(0), 475–478 (2013), In: Proceedings of the 2nd International Conference on Integrated Information (IC-ININFO 2012), Budapest, Hungary, 30 Aug–3 Sept 2012
8. Ganter, B., Wille, R.: Formal Concept Analysis: Mathematical Foundations, Springer, New York (2012)
9. Ge, X., Wang, P., Yun, Z.: The rough membership functions on four types of covering-based rough sets and their applications. Inf. Sci. **390**, 1–14 (2017)
10. Glodeanu, C.V.: Knowledge discovery in data sets with graded attributes. Int. J. Gen. Syst. **45**(2), 232–249 (2016)
11. Hill, J., Walkington, H., France, D.: Graduate attributes: implications for higher education practice and policy. J. Geogr. High. Educ. **40**(2), 155–163 (2016)
12. Ikram, A., Qamar, U.: Developing an expert system based on association rules and predicate logic for earthquake prediction. Knowl. Based Syst. **75**, 87–103 (2015)
13. Kianmehr, K., Alhajj, R.: Carsvm: a class association rule-based classification framework and its application to gene expression data. Artif. Intell. Med. **44**(1), 7–25 (2008)
14. Krajči, S.: A generalized concept lattice. Logic J. IGPL **13**(5), 543–550 (2005)
15. Kuhr, T., Vychodil, V.: Fuzzy logic programming reduced to reasoning with attribute implications. Fuzzy Sets Syst. **262**, 1–20 (2015)
16. Lu, H., Han, J., Feng, L.: Stock movement prediction and n-dimensional inter-transaction association rules. In: Proceedings of the ACM SIGMOD Workshop on Research Issues in Data Mining and Knowledge Discovery, p. 12 (1998)
17. Luxenburger, M.: Implications partielles dans un contexte. Mathématiques, Informatique et Sciences Humaines **29**(113), 35–55 (1991)
18. Malerba, D., Lisi, F.A., Sblendorio, F.: Mining spatial association rules in census data: a relational approach. In: Proceeding of the ECML/PKDD?02 workshop on Mining Official Data, University Printing House, Helsinki, Citeseer (2002)

19. Medina, J., Ojeda-Aciego, M., Ruiz-Calviño, J.: Formal concept analysis via multi-adjoint concept lattices. Fuzzy Sets Syst. **160**(2), 130–144 (2009)
20. Nahar, J., Imam, T., Tickle, K.S., Chen, Y.-P.P.: Association rule mining to detect factors which contribute to heart disease in males and females. Expert Syst. Appl. **40**(4), 1086–1093 (2013)
21. Ordonez, C., Santana, C., de Braal, L.: Discovering interesting association rules in medical data. In: Proceedings of ACM SIGMOD Workshop on Research Issues on Data Mining and Knowledge Discovery, pp. 78–85 (2000)
22. Paranjape-Voditel, P., Deshpande, U.: An association rule mining based stock market recommender system. In: 2011 Second International Conference on Emerging Applications of Information Technology (EAIT), pp. 21–24, Feb 2011
23. Popescu, A.: A general approach to fuzzy concepts. Math. Logic Q. **50**(3), 265–280 (2004)
24. Rai, S., Sharma, S.: Determining minimum spanning tree in an undirected weighted graph. In: 2015 International Conference on Advances in Computer Engineering and Applications (ICACEA), pp. 637–642, Mar 2015
25. Rudin, C., Letham, B., Salleb-Aouissi, A., Kogan, E., Madigan, D.: Sequential event prediction with association rules. In: 24th Annual Conference on Learning Theory (COLT 2011), pp. 615–634, July 2011
26. Vychodil, V.: Computing sets of graded attribute implications with witnessed non-redundancy (2015). arxiv: CoRRabs/1511.01640
27. Vychodil, V.: Rational fuzzy attribute logic (2015). arxiv: CoRRabs/1502.07326
28. Vychodil, V.: Computing sets of graded attribute implications with witnessed non-redundancy. Inf. Sci. **351**, 90–100 (2016)
29. Wang, W., Wang, Y., Bañares-Alcántara, R., Cui, Z., Coenen, F.: Application of classification association rule mining for mammalian mesenchymal stem cell differentiation. In: Perner, P. (ed.) Advances in Data Mining. Applications and Theoretical Aspects, Volume 5633 of Lecture Notes in Computer Science, pp. 51–61. Springer, Berlin Heidelberg (2009)
30. Yang, B., Hu, B.O.: On some types of fuzzy covering-based rough sets. Fuzzy Sets Syst. **312**, 36–65 (2017) (Theme: Fuzzy Rough Sets)
31. Zadeh, L.A.: Fuzzy logic. Computer **21**(4), 83–93 (1988)

Printed in the United States
By Bookmasters